Whispers of the Mind

Whispers of the Mind

A Neurologist's Memoir

CAROLYN LARKIN TAYLOR, MD

SHE WRITES PRESS

Published 2025
Printed in the United States of America
Print ISBN: 978-1-64742-936-2
E-ISBN: 978-1-64742-937-9
Library of Congress Control Number: 2025906237

For information, address:
She Writes Press
1569 Solano Ave #546
Berkeley, CA 94707

Interior Design by Andrea Reider

She Writes Press is a division of SparkPoint Studio, LLC.

Names and identifying characteristics have been changed to protect the privacy of certain individuals.

Dedication

This book is dedicated to the memory of my mother,
Jeannette Wix Larkin, my greatest inspiration
and source of inner strength.

Contents

Introduction xi

Preface xiii

"What's a Nice Girl Like You . . ."

PART ONE: MEDICAL SCHOOL—
CHOOSING THE UNEXPECTED PATH

Chapter 1

Rite of Passage 3

Chapter 2

Genetic Code Awry 10

Chapter 3

Anything Is Possible—Balancing Life as a Doctor and as a Mother 15

PART TWO: MEDICAL INTERNSHIP—
WHISPERS OF LEARNING AND LOSS

Chapter 4

You Can't Kill a Vet 29

Chapter 5

Code of Ethics 40

Chapter 6
Jenny's Story 47

Chapter 7
Mother's Day 52

PART THREE: NEUROLOGY RESIDENCY—
STEPPING INTO MY OWN

Chapter 8
Every Day Aboveground Is a Good Day 59

Chapter 9
Misdirected Guilt 68

Chapter 10
Right Brain, Left Brain 75

Chapter 11
Righteous Intolerance 79

Chapter 12
Grief Waits 85

PART FOUR: PRIVATE PRACTICE—
THE ART OF DIAGNOSIS

Chapter 13
The Soccer Mom's Double Life 93

Chapter 14
When Innocence Turns Deadly 100

Chapter 15
The Complicity of Parents 105

Chapter 16
Difficult Conversations 112
Chapter 17
The Invisible Thief 121
Chapter 18
The Case of the Misfolded Protein 132
Chapter 19
If Only 140
Chapter 20
Lewy Body Karma 145
Chapter 21
Death with Dignity: The Failed Video Recording 153
Chapter 22
A Tight Squeeze 163
Chapter 23
Sorry 169
Chapter 24
Prancer Tales 173
Chapter 25
Angels Among Us 177
Chapter 26
Please Don't Cry 182
Chapter 27
It Wasn't the MS 189

Chapter 28
Lightning Lessons 194

Chapter 29
Medical Gaslighting 200

Chapter 30
Thank You for Your Service 208

Conclusion 215

Acknowledgments 217

Introduction

Shh! Do you hear it? Can you hear the whispers?

I have heard them for as long as I've been a doctor. You learn to listen to the whispers that tell you what a patient is feeling, what is going on in their mind. They may not always say it clearly, so you must listen carefully.

Whispers of the Mind is a collection of essays that chronicles my journey and growth as both a neurologist and a human being. These essays are based on true stories I recorded spanning the course of my career through medical school, medical internship, neurology residency, and private practice.

I began this expedition as a coping mechanism, in a world in which doctors are trained to distance ourselves from our emotions to be more effective healers. Writing has enabled me to navigate the emotional difficulties entwined with the art of healing.

At the heart of this collection lies a journal—a sanctuary where I managed my own emotions as I witnessed the suffering of my patients, which at times mirrored my own personal struggles. This is my truth, as I learned much from my patients about what it means to be human, grappling with the physical and emotional challenges we all face in life.

Medicine is my true vocation. Neurology seemed to choose me along the way, calling to me in an otherworldly way that I had no choice but to follow. Neurology is not often a happy field, but it can be a profoundly rewarding one. The intricate tapestry of the human brain, combined with its profound connection to the heart and soul, makes it a fascinating field of study. More than that, it provided me with the unique opportunity to assist some extraordinary souls through their difficult journeys. As I diagnosed, treated, and guided them, I witnessed the resilience of the human spirit and learned profound lessons about life, death, and the beauty of human courage.

In sharing these essays, I hope to give you a glimpse into the mystique of the human mind and the resilience that lies within us all.

This memoir is a truthful recollection of actual events. To maintain patient anonymity, I have changed some identifying characteristics and details, such as descriptions and names of individuals and places.

PREFACE

"What's a Nice Girl Like You . . ."

We all know how that saying ends . . . "Doing in a Place Like This?"

Neurology is a field that most medical students avoid. There are GPs galore, but neurology has never been a favored field. It isn't a happy place. I rarely CURED anyone as a neurologist. I was often the one who diagnosed and managed the care of patients with often-terminal diseases.

I was the young mother of a beautiful little boy, treating diseases like Parkinson's, dementia, and ALS and managing stroke and brain tumors, as I was the one who brought those diagnoses and the prognoses to those patients.

And there I was, a young mother, telling people they were going to die.

Death is a journey for some, an event for others. It is the neurologist's job to help people on their journey. At least that was my perception as I made my way through each day. While there may seem to be a lot of talk about death in the essays in this book, death really isn't the point of all this. These are really essays about the courage and resilience of those facing it.

I started my career intent on becoming an ophthalmologist. Ophthalmology is a highly desirable specialty, and one that would have been a natural for me since I was already an optometrist. But when I rotated through the various specialties in medical school, I found myself being drawn to neurology. To me, it was just so much more interesting. There were so many more puzzles to solve, so much more detective work. And it offered so much more opportunity to connect with the lives of my patients. And that was what I loved most.

Neurology gave me the chance to use my talent as a diagnostician to the fullest. It gave me the chance to make living easier for those with neuro diseases, and to humanize the process of delivering "bad news" to patients in an empathetic way, to help minimize their fears, and to let them know that I was with them through it all.

Through my journey in neurology, I came to understand that the real heart of medicine lies not only in diagnosing or treating diseases but in walking alongside my patients as they face their most challenging moments. Even when cures are not possible, compassion always is. And in the end, that is what sustains both patient and doctor. I learned to embrace not just the science of neurology, but the human stories behind each case—the courage, the fear, the resilience.

So, what was a nice girl like me doing in a place like this? I was bearing witness to life in all its fragility and strength, offering a hand, a word, a heart—helping my patients find dignity and grace, even in the face of terminal illness. Because, in the end, it wasn't just about death; it was about living fully, right until the very last moment.

PART ONE

Medical School

Choosing the Unexpected Path

CHAPTER 1

Rite of Passage

Like the closing chapter of a well-loved book, death beckons us all with an air of finality and mystery, leaving behind the lingering echoes of stories untold. The only thing in life that is certain and inevitable, death binds us all together with this one common destiny. It is the one truth that unites every living creature on this earth. No matter who we are or where we come from, death will come to us all.

My first experience of seeing a dead body was at my grandfather's wake. I was nine years old at the time and was given the option to wait outside. Curious, I was unable to, and so I ventured in, joining the somber line of adults to walk by his open casket. I recall how very peaceful he looked, dressed in a black suit and tie, his countenance serene, his eyes closed. I imagined that he was still there sleeping, somehow aware of the many visitors who flocked there to pay their last respects. As I knelt by his coffin, I silently told him how I had overheard my grandmother speaking about how terrified he had been the night he died. As a physician himself, he must have known he was going to die that night and was afraid to lie down, almost as if sitting up would fend off the inevitable.

Medical school is a place where we learn how to heal and comfort the living, how to honor and protect life. But life and death are so inextricably entwined that death must also play a significant role in our education. And so, we are also in the business of helping those whom we can no longer heal to accept this inevitability, of easing the ordeal ahead of them. This is a profoundly difficult task, for we're not immune to this fate ourselves, and to aid our patient we must also accept this as our fate at some undisclosed time. It is sometimes interesting to observe the temperaments of medical students faced with this dark task, perhaps guiding their choice of specialties. There are definite dermatology personalities, oncology personalities, surgical personalities—for their choice will determine the types of cases they will assist for the remainder of their working life.

As a first-year medical student, I was required to take gross anatomy, which has an obligatory lab course where we dissect a human cadaver. In the 19th century, they sometimes robbed graves to obtain cadavers for dissection. Today, cadavers are obtained from unclaimed bodies or those who will their bodies to research, but there are requirements that these willed bodies must not be too obese or riddled with too much disease, so there have been declining numbers.

Because donated bodies were becoming a rarity worldwide at the time I entered medical school, there were three students assigned to each cadaver. Our instructor began the first session with a standard speech. "Today," he said, "marks a rite of passage for every medical student. Today, you have the enormous privilege of dissecting a human body. You are to respect these bodies as they were once people like you who have generously donated their corpses for the sole purpose of your education. Over the course of this semester, you will get used to the putrid smell of formaldehyde along with the lingering smell on your clothes when you leave. As nauseated as many of you may be right now, by exam time you will all be putting in late hours with a scalpel in one hand and a sandwich in the other. Good luck."

My cadaver that day was an elderly woman. We chose not to give her a name to make it easier to get through the task at hand. In life she could not have weighed more than one hundred pounds, so I assumed she suffered from either protein calorie malnutrition from neglect or possibly cancer-related cachexia. Her gray hair was matted to her scalp, her mouth ajar and body rigid. Her eyes, the windows to the soul, were open and vacant with that glazed, gray discoloration typical of the dead, once the soul has vacated its home. These bodies were not embalmed, so they presented as they were at the moment of death.

I drew the short straw, so I was to be the one to use the electric saw to cut through the sternum to open her chest, and through her skull to expose a tiny, shriveled brain. I put on goggles to shield my eyes from the bone fragments as the saw busily buzzed through the hard bone. Heart, lungs, stomach, liver, spleen, pancreas—all were exposed for us to dissect and study. We carefully mapped out the course of every major blood vessel and nerve, inserting pins as markers of points we wanted to return to and study later.

The day we dissected the bowel was particularly gruesome, with a stench that not even formaldehyde could mask.

We traced the miles of bowel from stomach to rectum. There it was, a large cancerous tumor, its runaway cells grown out of control, the cause of her demise. We lost a few students that semester to loss of consciousness from fainting, but all save one returned to dissect another day. And confirming our instructor's prediction, by midsemester we found ourselves studying our dissections for our exam complete with popcorn, pizza, and coffee—lots of coffee—without a thought about it.

Pathology, the study of disease states, was more interesting to me, as I would be able to figure out what was wrong with my patient. I particularly liked studying histology slides, visualizing cancerous cells gone awry, taking over like a great white shark consuming all the schools of smaller fish in one large gulp. Somehow, cells on a slide seemed less real, and appeared to have the potential to be managed.

An autopsy was a critical part of any pathology course, but with fewer families consenting to these now, if it wasn't a coroner case, we might not actually have gotten the chance. The autopsy rate in hospitals had been declining worldwide for decades. With improved diagnostic capabilities determining cause of death, there was very little reason for families to desecrate the body of their loved one postmortem for answers anymore. We were nevertheless all given a beeper just in case a body should arrive in the middle of the night or weekend in the hope we'd get that chance.

Early one morning while I was in class, my pager went off, signaling that I was to report to the city morgue right away. Upon arrival I was ushered in, and the familiar nausea from the smell of formaldehyde brought the gross anatomy lab roaring back to my senses. Lying naked on a slab of steel was a nine-year-old girl who had died suddenly from complications of acute lymphoblastic leukemia (ALL). It was one thing to dissect an old woman who had lived her life, quite another to desecrate the pristine body of this beautiful child paused on the very cusp of puberty—a life snuffed out by cancer before it had even begun. Such a cruel and indiscriminate blight upon life. Her body was still so perfect, with remnants of peeling pink nail polish on her fingernails and blonde curls framing her angelic face. My heart cried out silently as I watched the pathologist wield his shiny steel scalpel in a T-shaped pattern, first down the length of her torso and then just below the sternum, so that her interior cavities were laid open for all to see.

Her organs—liver, spleen, heart, and lungs—were still so pink, glistening and healthy, without the telltale signs of age and disrespect so often seen in the elderly lifelong smoker with blackened lungs and calcified blood vessels. Next came the dissection of the great vessels, with their youthful suppleness and clear, glistening interior walls unmarked by cholesterol and calcium deposits from life and stress. This is how our bodies are meant to be, the way we all start out. How can something as perfect as this succumb to the ravages of cancer? Her leukemia, visible only under the microscope, had presented so rapidly that she died before treatment could even be initiated.

The whole time I watched the dissection of this child, I tried to remind myself that this was a valuable learning opportunity that I dare not waste and I compelled myself to concentrate as hard as I could on the anatomy and pathology at hand rather than what she might have been doing last week.

The autopsy was now complete. Her organs were in jars for further histological study and her torso was closed with sutures, ready for her visit to the mortician who would bathe, dress, and apply enough cosmetics to make her look almost alive for viewing by her loved ones before she was laid to rest.

As I was gratefully about to leave, I was called over to another steel slab by one of the morticians. "Hey, this is your lucky day. We have another case for you, one you may never get to see again."

Just as I thought this couldn't possibly be any worse than the last one, I saw a small, incompletely formed infant, a late-term fetus. A therapeutic abortion had been performed for a severe chromosomal abnormality called Turner's syndrome, which I'd only read about before this. This child would have been a girl, but instead of the XX female chromosome, she'd lost one X and was XO. She was neither male nor female, with no genital organs. Instead of appearing normal on the outside, she had a disfigured webbed neck.

As I watched the second autopsy, I was filled with conflict. If this were my child, what would I have done? I had not even thought about having a family yet, although I supposed I would someday. I had never actually contemplated how I felt about abortion morally, but perhaps it was time I should. This child, if brought to term, would have had to face so many challenges: disfigurement, lack of primary and secondary sexual characteristics, low intelligence. Many chromosomal abnormalities are incompatible with life and self-abort, making the choice for the parents. How does a parent bring themselves to make this choice? Contrasting this planned demise of a child that appeared so imperfect with the tragic loss of the perfect-appearing nine-year-old girl from leukemia on the next metal slab, I couldn't begin to imagine the agony both parents must have gone through, and I hoped that I would never have to make that choice.

Too late to return to class, I stopped at the university pool—a place that always helped me relax. As I crossed the pool doing lap after lap, I found myself unable to empty my mind in the meditative way that had always worked for me. Instead, each stroke and deep breath seemed to recall the vivid images of the two young bodies I'd witnessed earlier, their torsos cut open, revealing the empty cavities where their perfect organs were now removed. It was a haunting sight, and I couldn't help but think that these children had been alive just twenty-four hours earlier.

CHAPTER 2

Genetic Code Awry

I first met Sam during my medical school neurology rotation. A thirty-two-year-old man, he had been referred to our clinic because of involuntary movements. His family history was unknown as he was adopted. He had been normal growing up, finished college, and had a good job until about two years before he presented to our clinic, when his personality started to change. His father related that Sam started drinking and partying excessively and his behavior had become erratic. His cognitive functioning also appeared to be declining, so he had to move back home as he was no longer able to hold down a job. Initially seen by psychiatry with the suspicion of bipolar disease, it was not until he started displaying involuntary writhing movements that it occurred to his doctor that this might be a neurological condition. Medical school clinics were generally open to the needy for free. Sam had lost his health insurance when he lost his job. So, in exchange for free care, Sam was being used as a teaching case for the medical students.

Sam and I were close in age, and my heart went out to him as I watched him moving uncontrollably, with limbs flailing and contorting as if under the control of an invisible puppeteer, as he was instructed to walk the length of the hallway so we could observe him. As he slowly ambulated, trying not to fall, his head, face, trunk, and limbs would suddenly jerk wildly, causing him to lose his balance. When he tried to stand still, his limbs would display a continuous writhing type of movement, giving away his lack of control. He must have been so self-conscious in front of all these people, like a puppy in the window of a pet store, but he was willing to do anything if only we could make the proper diagnosis to help him. Our attending neurologist talked about his case in front of him almost as if he couldn't hear or understand while the students stood around eyeing him curiously. My heart heavy with sorrow, I wanted to shout out, "He can hear us, and he has feelings!"

The attending neurologist knew right away what this was and ordered the appropriate genetic testing. It confirmed a trinucleotide repeat in the HD gene, a genetic mutation in which a particular three nucleotide sequence (C-A-G) in the DNA is repeated multiple times in a row. Once the mutation occurs, it can be passed down from generation to generation in a dominantly inherited fashion, such that an individual has a 50 percent chance of inheriting the repeated sequence. The number of repeats also tends to increase with each generation. Once the number of repeats exceeds a certain threshold, it can lead to the development of a disease, in this case Huntington's chorea.

Huntington's chorea is a relentless and devastating disorder, marked by the gradual loss of control over both body and mind. In its early stages, the disease whispers through subtle tremors, barely noticeable at first. But as it progresses, these movements grow more erratic, with involuntary jerks and spasms that disrupt even the simplest actions. A once steady gait becomes unsteady, each step uncertain. Yet, the physical decline pales in comparison to the inner toll. Huntington's slowly strips away the essence of self, unraveling cognition and clouding memory. What was once a clear mind becomes tangled in confusion, as the disease fractures the very core of the person.

The slow progression is particularly painful for both patient and family. Perhaps the cruelest part of this is the inherited nature of the disease, which might explain why Sam was given up for adoption. It is particularly tragic when the diagnosis is not made until after an individual has already had children, sentencing them to this terrible uncertainty. There is no known cure.

And yet, despite the severity of this disease, I have witnessed incredible moments of love and courage. Brian was just twenty-eight when he came under my care many years later in private practice. He had watched his mother and oldest brother succumb to and die of the disease. His second older brother had already been institutionalized. His father brought him to me, looking for any help he could get. They had all been born before their mother manifested any symptoms or even knew she could have the disease herself. Brian, who grew up witnessing all of this, was very accepting of his fate. He was an incredibly gentle soul who volunteered to help others at the local shelter once a week with his father.

When I first met Brian, he had only minimal involuntary movements and was still able to drive and surf, his favorite pastime. Within a year, his father had to move him out of the apartment he shared with friends into a group home for the disabled, because his friends were taking terrible advantage of him, even stealing from him. Over time he lost the ability to do all the things he loved and went from being ambulatory to wheelchair-bound and finally unable even to speak. He was so grateful when I was able to get him on a new drug that reduced his choreiform movements substantially for a few years. His father would take him for trips to the ocean in his wheelchair so he could watch others surf, remembering how much he had loved it. As badly as I felt for Brian, I could not begin to imagine the pain his father must have lived through, not only losing his wife to such a horrible fate but having to watch all three of his sons confront the same terrible end.

One of the things I will remember most about Brian was the huge hug he would give me at the end of every visit. Brian taught me a lot in the years I cared for him. Amidst his uncontrollable movements, he demonstrated a fleeting grace, revealing his humanity even in the throes of such terrible despair. These glimpses of his spirit serve as a reminder of the resilience that endures within the hearts of those touched by Huntington's chorea. He taught me that we all have a journey that we are meant to take in this world, and we need to be accepting and even grateful that we are here in the hope that we can touch the lives of others as he touched mine in such a positive way. He had so much to be bitter and angry about, yet he always had a beaming smile for everyone. His attitude seemed to be that with so little time on this earth, he could not afford to waste it in bitterness. Instead, he spread joy wherever he went.

In the face of such profound suffering, I have been privileged to have witnessed such courage in so many of my patients. Huntington's chorea, with its random attacks on young life, reveals the profound fragility of the human experience, and should remind us all of our delicate mortality.

CHAPTER 3

Anything Is Possible

Balancing Life as a Doctor and as a Mother

When I first told my mother that I wanted to be a doctor, her response wasn't what I expected.

"Don't you want a family someday?" she asked. "Being a doctor is a hard life. You'll miss so many moments—moments you can't get back."

As the daughter of a physician, she knew what she was talking about. I nodded, thinking I understood. But the truth is, I didn't. My ambition to become a doctor drowned out her words of caution. Back then, I hadn't given much thought to having a family—I believed I could make it all work. Now, looking back, I realize how naive that belief was.

And the panic that gripped me when I learned I was pregnant in my third year of medical school.

We were on student hospital rounds with one of the cardiology attendings. He was instructing us to rely on our knowledge of what felt normal to us in terms of heart rate at the patient's bedside instead of looking at the second hand on our watches. Using me as an example, he said, "Check your own pulse first, then check the pulse of your patient, and you should know right away if the patient's pulse is too fast or slow in comparison to your own." So, I did just that and competently stated that the patient's pulse was much too slow.

"Impossible," he said as he reached over and checked my pulse. "There's something wrong with you. Your pulse is over 180!"

With that, I suddenly felt faint and slumped down to the floor. Recovering quickly, I later realized I hadn't eaten that morning and once I got some food in my system, I felt much better.

This event set off an alarm in me. Something was amiss. I could feel it.

The thought of pregnancy crossed my mind, but I delayed taking a pregnancy test, fearing the results. If it was positive, I was afraid the news would deter my preparation for a big exam I was getting ready to take. I had my yearly gynecology exam scheduled for the following week, and I decided finding out if I was going to be a mother or not could wait.

And then it came.

"Congratulations! You're pregnant!" my GYN proclaimed. "Your estimated date of confinement is December 7."

I can still hear those words echoing in my memory. He used the medical term "confinement" instead of delivery, which summed up my fear of how I was going to feel—confined. I was thirty-three at the time and had been married for eight years. My husband and I never decided not to have children, but then we never decided to have them either. I guess this child we had conceived was the one who decided for us.

My husband, Ken, was truly delighted, popping a bottle of champagne to celebrate. The mere thought of drinking made me queasy, not to mention the fact that it was not good for the baby. Baby! I was having a baby! I was honestly more terrified than thrilled at first. I recall thinking that, from now on, I was responsible for another life, at least for the next eighteen years. Little did I realize at the time that eighteen years was a drop in the bucket. I still worry about that child who is now a grown man in his thirties!

My mind raced about all the things that could possibly go wrong. What if I had difficulty with the pregnancy? What if my child was not healthy or had a genetic abnormality? What if I couldn't arrange reliable childcare, or what if my child were sick? How could I bring myself to leave this baby with just anyone? How was I ever going to manage the grueling hours of internship and residency if I was responsible for a young child at home that needed me?

Ken was wonderful and assured me at every turn that somehow everything would work out. Between the two of us, we would find a way. Neither of us had family close enough to help so we were truly on our own.

As it turned out, my pregnancy was uneventful except for the usual fatigue and discomfort all women experience. And I was engaged in hospital rotations, sometimes working as much as eighty hours a week and taking overnight calls. There were days when my ankles were so swollen that I couldn't get my street shoes back on after being up working all night.

Times were different then. It was the '80s, a time when far fewer women were in medical school and nothing was in place to accommodate a pregnant student, such as maternity leave. I had the option of taking a leave of absence for a year or just plodding on. I was afraid if I left that I would never return, so I chose to plod on.

I managed to schedule a "research rotation"—a month of independent research without actual patient duties—around the time my baby boy was due. Completing my OB rotation two days before Thanksgiving, I was exhausted and ready for a break. My baby was due in just ten days, and with any luck he would be on time. With my surgery sub-internship beginning in just one month, I hoped and prayed for a sooner rather than later delivery so I would have a little more time to bond with my child.

Rory cooperated by deciding to arrive early, and I went into labor with him the following day. He was born on Thanksgiving, a healthy seven-pound baby boy. That was the happiest moment of both his parents' lives. The love I felt for my child was indescribable. I didn't know how I was ever going to leave his side to complete my training. I finally had the first inkling of what my mother had cautioned me about, and this was just the beginning.

Motherhood isn't all hearts and flowers, despite what people make it out to be. In those first few weeks, I was exhausted all the time. I tried breastfeeding, but Rory threw up almost everything he ate. He was constantly hungry and screamed nearly twenty-four hours a day. Frustrated and desperate, I once sneaked into the basement, stuck my head in the dryer, and screamed as loud as I could, just to release the tension.

I remember thinking that even my thirty-six-hour hospital shifts were easier—at least then I had a better chance of getting some sleep.

I stuck with the routine of Rory feeding, throwing up, and crying for as long as I could. Finally, the pediatrician suggested switching to formula feeding. I thought it would be the solution, but the same cycle continued. Rory still drank just a little and vomited everything up. He was starving and not gaining weight as he should. By the time he was three weeks old, his vomiting became projectile—so severe that it sometimes hit the bathroom wall when I tried to burp him. We resorted to feeding him only in the bathroom because it was easier to clean up. I kept calling the pediatrician, but he brushed me off, treating me like an anxious new mom who happened to be a medical student, overreacting. It wasn't until Rory's vomiting became projectile, a sign of a more serious obstruction, that he finally listened.

After a barium X-ray, Rory was diagnosed with pyloric stenosis—a condition where the muscle controlling the passage from the stomach to the intestine thickens, causing a blockage. He had surgery that same day to fix it. At almost four weeks old, Rory was finally a normal baby—eating, sleeping, and cooing. Within two weeks of the surgery, he had gone from the 50th to the 100th percentile in weight. To this day, he's still hungry all the time, eating every couple of hours. I don't think his body ever forgot those early weeks of hunger.

Just as we got through the crisis, I had to return to the hospital for my surgical sub-internship, which included overnight shifts every third night. I didn't know how I was going to bear it.

Now that Rory was happily taking formula, Ken could help with the feeding. I would get up at 4:00 a.m. for the first feeding, then hand Rory off to Ken, who would take care of him until the nanny arrived. I had to leave by 4:30 a.m. for my ninety-minute drive to the hospital—an assignment I had to take when I found out I was pregnant because I had to readjust my schedule, and this was the only one available. I would arrive by 6:00 a.m., round on the patients, handle basic tasks like drawing blood and changing dressings, and then assist or observe surgeries while getting quizzed by the attendings on anatomy and procedures. My day usually ended around 6:00 p.m. after postsurgical rounds, but then we had to attend the chairman's rounds, which kept us there until 8:00 p.m. The earliest I ever got home was 9:30 or 10:00 p.m., and that was only two out of every three nights.

I thought I was going to die.

I hated every minute away from my newborn, especially knowing he was with a nanny I didn't know well whenever Ken was at work. I don't know who needed the other more—me or my baby.

I had to do something to fix the situation, but what?

Unable to bring myself to outright leave, I came up with a plan. The next day, I boldly marched into the surgical chairman's office. He was an intimidating figure of a man—tall, overweight, balding, with black rimmed glasses hanging low on his nose. All the residents and medical students feared him, as he wielded all the power and barked all the orders.

"Dr. B., I have something to tell you," I began. "I will be the best medical student you have ever had rotating through your department. I will be on time, work hard, and take call, but *I will not* stay for your required surgical conferences in the evening."

"Is that so?" he said as he peered at this female medical student in front of him. "And just who do you think you are to dictate the terms of this rotation?"

I was so hoping that he would just tell me to leave the rotation, which was exactly what I wanted as I couldn't bear it any longer. So I answered honestly. "I have a newborn infant at home, which is a ninety-minute drive from this hospital, and I barely get any time with him as it is."

I was so sure—so sure—that was going to be it. How dare a young medical student—and a woman at that—commit the cardinal sin of insubordination before such an imposing personality.

To my surprise, he answered, "All right, then, but I'll hold you to your word about your performance."

I was doomed to stay, but at least I got home a little earlier to spend some time with my little boy. And the surgery rotation was just six weeks, after which my next scheduled rotations were closer to home.

At the end of the rotation, I was called into Dr. B.'s office to discuss my performance. After the stunt I had tried to pull, I was certain I would get a poor grade. To my surprise, the conversation went very differently.

"Young lady," he said, "you have earned a high honors in this rotation. Excellent performance. I hope you're planning on going into surgery as you have a surgeon's personality and were unquestionably born to do this."

You could have knocked me over with a feather!

And so it went. Rotation after rotation, with my husband's help and a series of wonderful European au pairs that lived in our home, I managed to graduate on time. Now I was off to start my internship, followed by a three-year neurology residency.

I can do this. I know I can. The words resonated in my mind.

There were so many nights when I had to stay in the hospital to take call. Those were the nights that I missed my little boy the most. I yearned to be the one giving him his bath and telling him his nighttime stories. I would call home and try to talk to him on the phone, asking him how to make the animal sounds—"How does the dog bark? How does the bear growl?" When I would hang up, I often cried, if I could find a quiet place where no one could see me. I recall calling my mother many nights tearfully relating how I didn't think I could go on because I missed Rory so much. I still recall her words to this day.

"Rory is just fine. You're the one who is suffering right now and that's only natural. You must keep going. You have come so far, worked so hard, and wanted this for so long. You need Rory now, but he is going to need you so much more as he gets older, and you will be a practicing physician by then. You will have more time and the means to help him when he needs you the most. And trust me, he will."

Oh, what a wise woman she was, my mother.

Each year of training got a little easier, or perhaps I was just getting more used to it. My husband would often bring Rory to the hospital on weekends when I was on call to see me if I could steal away for a brief time. Rory was so adorable, with his blond head of curls and blue eyes, as he would run into my arms declaring, "Mommy at *hopsital*." When I came home after a thirty-six-hour shift, he would be upset with me for having been gone so long. At first, he would refuse to talk to me but that didn't last long as we played Candyland together into the night. Or he would put on his Big Bird sunglasses and red galoshes and dance around the room in his blue diapers. The next day I would be frantic, looking for my beeper only to find that Rory had hidden it under his bed. "Mommy not go *hopsital* if she doesn't have the beeper."

Oh, my heart!

Things were not flawless, even with help at home from Ken and the au pairs. There was the day I walked out of a first-year neurology in-service examination because my three-year-old had developed an orbital cellulitis—a serious condition that can potentially develop into meningitis—and I had to take him to the hospital. My chairman at the time refused to allow me to leave, stating that someone else would have to take care of it for me. Clearly, he was not a mother. My lack of performance on that test I walked out on was to have serious repercussions later.

There were so many precious moments I missed—the preschool Halloween parades and teddy bear picnics, Easter mornings, class trips, school plays, kindergarten soccer, swim meets, and so many little moments of just spending time watching my toddler play. I always felt tired, trying to pry my eyes open and drinking tons of coffee to stay awake, just to be with him after being up all night.

Once training was over, I wound up working full-time in a private group practice, but it was so much more manageable.

I could schedule myself out for a special occasion—a play, a Little League game or special competition, or just to take Rory to a friend's birthday party—as long as I wasn't on call for my group at the hospital. It became normal for Rory to tag along with me if I got called in from home to the hospital for a consult and his dad was not around to help. The ER staff knew him well. I would get Rory settled into the ER doctor's lounge and they would put children's programs on TV for him, or we would bring his coloring books. He also became adept at amusing himself in the waiting room of the ICU.

In private practice I managed to work a four-day week when I wasn't on call. And that extra day was "our special day." Rory would get to decide what we would do together. I would pick him up after school like any other mom and we would go out for ice cream, perhaps go to the park, or to the store to buy a new Beanie Baby, or have a friend over while I baked cookies.

The best day was when he presented me with an ad he had circled in the classified section of the newspaper for a golden retriever puppy that he wanted to look at. It was his day to choose what we would do, and even though I told him I was not going to buy a puppy, I knew enough to bring my checkbook. Who has ever been able to resist a golden retriever puppy? As serendipity would have it, there was just one puppy left and her name was Rory, short for Aurora. How could I say no?

Those days felt so normal! And I so cherished every moment. There were times I felt like I had more quality time with my son because we always made the most of what little time we had together.

Rory turned out to be a gifted athlete at almost every sport he tried. He was skiing the expert slopes by the time he was four, played soccer, was on the swim team, and later excelled as a baseball pitcher and football quarterback. I was able to attend most of his games, unless I was on call. And I was there for the broken ankle, broken leg, and two broken noses.

True to my sweet mother's prediction, Rory did need me much more as he grew into his teenage and young adult years. And as a physician with a career, I was in a much better position to help him. How did she know?

When I ask my son today if he feels he missed out on anything as a young child since I was away in training so much, his standard reply is, "I don't know what you're talking about, Mom. I had a great childhood."

My advice to any new mother contemplating a career in medicine is simple. *If medicine is truly what you really want to do, what you need to do, then go for it! You will have to make sacrifices, but with a lot of love, anything is possible.*

PART TWO

Medical Internship

Whispers of Learning and Loss

CHAPTER 4

You Can't Kill a Vet

Internship year strips away any illusions, a trial by fire that every medical student both dreads and anticipates. For four years, I had braced myself for this moment, imagining it long before I ever donned a long white coat. No longer would there be an instructor hovering close by, ready to step in if something went wrong. The safety net would be gone, leaving only me, my instincts, and my knowledge. To prepare, I'd pushed myself to the edge during my final year of medical school, choosing the most grueling electives, chasing every challenge. The mantra was simple: "See one, do one, teach one." Whether it was threading a central line into the neck or piercing the spine for a lumbar puncture, there was no hand-holding, only a fleeting chance to observe before it became my turn. And when that moment came, I had no choice but to get it right.

I was assigned to the Veterans hospital for my first of three monthly rotations required during my internship year. I had heard this was the most dreaded rotation because of the lack of ancillary services at the VA hospital. I wasn't sure exactly what that meant, but as far as I was concerned, if I was going to have to get my feet wet, it might as well be there. The only comforting words that I heard from residents who had gone there before me were, "Not to worry. No matter what you do wrong, you can't kill a vet." That was the thinking. They'd survived the horrors of war, far worse than any of us could possibly inflict on them because of our inexperience and incompetence. So we would be safe with them, regardless of our lack of experience.

My first day finally arrived when I got to trade in my short white student coat identifying me as a student for my newly acquired coveted long white coat proclaiming real doctor status. We were "on call," meaning we worked a thirty-six-hour shift every third to fourth night, and I was assigned to take my first call on my very first day, July 1. Hopefully, I would manage to get some sleep during the night if it wasn't terribly busy. I'd never worked in a VA hospital before, but I had heard stories! Trying to remain calm about having to take call on my first day, I reassured myself that it would provide me with a crash course in how the hospital worked, and after my first rotation there, it should be smooth sailing.

I arrived early that first day, to meet my resident on the third floor along with five other new interns from various medical programs. Dr. R. was an impressive fellow, standing tall in his freshly starched white coat, peering over his glasses at his frightened new flock of interns. We were a diverse group, representing three different Philadelphia medical schools, gathered here to begin our careers. We were two women and four men eagerly awaiting our assignments.

Older than the others, including my resident, I immediately felt a little out of place. I was also conscious of the fact that I was probably the only mother in the group, my eighteen-month-old son never far from my thoughts. *How will I ever manage the long hours and overnight calls away from him? What if he gets sick? What if I just can't bear to be away from him for the necessary four years of training I signed up for? How will I manage if my carefully planned day-care arrangements fall through?* There were no allowances for such things in medical training. If one of us was absent or late, that put added pressure and work hours on the entire team who had to take up our slack. I imagined these were just some of the reasons women might face discrimination during the application process, and I was determined to prove that I could perform just as well and work just as hard as any of the men.

Wasting no time, Dr. R. began. "Looks like everyone's here," he said, his eyes surveying the eager group of newcomers. "Welcome to the VA. I know you've all heard a lot of stories about the difficulty of working here, and they're probably all true. But you'll all get through it and be better doctors from the experience."

We proceeded with our patient rounds, going from bed to bed, as Dr. R. introduced each patient by age, admission date, and symptoms, pointing out all the salient features on examination as he grilled each of us about the differential diagnosis and testing we should pursue. Dr. R. randomly assigned each patient to one of us to manage since the June group of interns had gone off service.

As the intern on call that evening, I was responsible for all new admissions coming through the ER over the next twenty-four hours and my beeper had already started to go off repeatedly during our patient rounds. Each time it went off, I excused myself from the group to take the call, and each time I was informed by the ER that there was a new admission waiting there for me. I acknowledged the notifications as they came, wondering why there were so many so early and I had been told none of them were urgent. I would later learn that it was a common occurrence right before a holiday weekend for chronically ill veterans to be dropped off at the VA ER in hopes of them being admitted so their families could get out of town for respite. Ah! This was beginning to make sense, as the Fourth of July holiday was coming up. Since none of these cases were urgent, I figured I could continue with patient rounds and wait until the new admissions had been transported to their assigned rooms and the nursing staff had gotten them settled in before catching up with them to perform their histories and physicals and write their orders.

Nearly three hours later, as teaching rounds finally ended, I found myself worrying about how much I still had to do—i.e., admitting four new ER patients, taking over my newly assigned established patients, learning where everything was, how the hospital worked, etc. *Surely this hospital works like every other hospital I've rotated through, right?* Wrong. Oh so wrong! I'll never forget the resident's parting words to us that day: "Whatever you need, whatever you do, do not call me, especially in the middle of the night, unless someone is actually dying." *Not me,* I said to myself. *I would never do that; I'm going to handle this.* All I needed was a pair of roller blades.

As it turned out, the chaos was about to begin. There was only one computer on each floor—all antiquated devices—and I couldn't get anywhere near one. Frantically, I tried locating the charts of my assigned patients after examining them, without success. In the past I'd always been able to rely on the nursing staff for help, but they seemed to be scarce here, and when I did manage to find one, she barely spoke English. Then, to my shock, I discovered that my patients' vital signs (temperature, heart rate, blood pressure), taken by the nursing staff and posted outside their rooms that day, were all identical—recorded, but not actually checked! I realized this when, to my horror, I entered one of my patients' rooms and found him in rigor mortis, indicating he had been dead for many hours despite the normal vital sign reading posted outside the door, and with no heart monitors in the rooms. When I finally gained access to the computer to check all my patients' blood work results, I discovered that many of these results were missing. I soon realized that if there was not a fresh cotton ball taped to a patient's arm that day, the blood had not been drawn. When I tried to reorder the blood work, I was informed that if I wanted blood drawn after the 6:00 a.m. blood draw, I must draw it myself. As hard as I tried to remain calm, I was now beginning to panic. I now had eight new admissions waiting for me, and it wasn't even nightfall. In addition to managing the patients on the ward I inherited, plus covering for all the other interns' patients after they left for the day, how was I ever going to manage these new admissions—history, physical, and orders—in only twenty-four hours?

Just when I thought things couldn't possibly get any worse, I received another page from the ER, this one from an angry nurse.

"Exactly when are you planning on coming down to retrieve your admissions?" she yelled. "They're all lined up in the hallway waiting for you and we're running out of room."

"What do you mean lined up? I was waiting for them up on the floor."

"You must be dreaming," she yelled back. "You're an intern. In this hospital it's your job to transport your patients up to their rooms, get them into bed, establish any IV lines, write the orders, draw any necessary blood work, etc. Now get down here."

Taking a deep breath, I finally realized just why this rotation was so dreaded and what they meant by poor ancillary services. Welcome to the VA, where there are no transport services, IV teams, or hematology for blood draws. Just interns! Now I understood the sign above the door to the hospital, which read, "God bless our patients and God help our doctors." As I worked throughout the night, admitting one patient after another, Dr. R.'s warning not to awaken him unless it was a matter of life or death echoed in my head.

With the other interns having wearily signed out to me by early evening, giving me a rundown on their patients in the event that there was a problem during the night, everything now fell in my hands. Sometimes there were late labs to check, IV fluids to adjust, or just the need for me to look in on them to see if they were unstable in any way.

Throughout the night I was called to evaluate every fever, which prompted drawing blood to check for infection, obtaining a chest X-ray and urine analysis to find the source of the fever, and treating with antibiotics if necessary. Since the turnaround time on lab results was slow, I found myself looking at slides under a microscope to see if I could identify any organisms indicating infection.

Since there were no IV teams, I was called to reestablish an IV line for anyone whose line came out. I recall one particularly difficult situation involving an elderly man on chemotherapy for cancer. His veins were so scarred from prior treatment that it was nearly impossible to get the IV established, and he was due for another round of chemo that night. I tried so many times to introduce the line, and each time the vein would collapse before I could get the fluid to flow. By about the tenth time, the patient was near tears. Despite this, I could tell he felt sorry for this newbie intern—i.e., me—struggling to get the line started in the late hours of the night. Finally, the lone nurse on his floor stepped up and offered to try and got it right away. I was so grateful!

Then I got what I thought would be my easiest call of the night. One of the patients being fed through a gastric tube had a blockage. In any other hospital, I would page the surgery resident on call and he or she would remove the blockage. But wait—this was not just any hospital. When I called him, he was sympathetic to my plight, realizing this was my first night there. What he said shocked me. "This is the VA so I'm going to tell you what you need to do here. There's a soda machine on the third floor. Get a can of Coke and pour it through his tube. It's stronger than Drano. That should do the trick." And it did.

In the early hours of the morning, I found myself in a large ward they called the ballroom. It was lined on each side by twelve single hospital beds, each holding one of our veterans, and separated by white cotton curtains softly blowing in the breeze generated by the large fans trying to cool them in this non-air-conditioned facility. As the sun peeked just above the horizon through the single smoke-tinged window, I was reminded of my place here. I was here to help take care of our veterans, to give them a sense of comfort and assurance that their needs were being taken care of with all the compassion and respect they deserve.

I had just enough time before the morning teaching rounds to process my last ER admission from the night before when the reverie of my fatigue was broken by my pager again. This call was from the holding area, where my patient was still waiting for me to arrive.

"Doctor, your patient is complaining of chest pain," the voice on the other end of the line said. "What do you want to do?"

What I really wanted to say at that moment was, "Well, hell . . . I don't know. I haven't even met him yet, and I know nothing about his medical history except that he is next on my list to admit." Instead, forcing my voice to a gentle calm, I said what I would say in any other hospital: "Order a stat EKG and I'll be right there."

The snickering voice on the other end of the line was now starting to sound familiarly irritating. "This is the VA. There are no EKG technicians. You're it. There is one EKG machine in this hospital. Find it and do the EKG yourself."

I had been hearing this familiar refrain for twenty-four hours now. There was of course no EKG tech, and there was of course only one EKG machine in this f–ing hospital! It was up to me to find it, bring it to the patient, and perform the EKG—provided, that is, I even recalled how to do this, since I had not actually hooked someone up to one of these machines since physiology lab in medical school. All this while my patient might be having a heart attack! There is nothing like knowing that you're the only one responsible, the only one who can help someone in need, to put you into absolute fear gear.

I raced frantically from floor to floor until I found the EKG machine, then wheeled it as fast as my weary legs would carry me to the patient. When I finally entered the holding area, I was ushered in to see a man in his late sixties lying on a gurney and covered in a white cotton blanket. He had an oxygen mask on his face, which was pale and drenched in sweat, his breath came in gasps, and he clutched his chest in pain.

After assessing his vitals, I ordered a sublingual nitroglycerin to buy myself some time while I struggled to hook up the EKG leads, first to his chest and finally to his legs. When I pulled down the blanket, I was dismayed to find two stumps where his legs were supposed to be—another surprise! I struggled to hook up the limb leads somewhere on the stumps, then took a deep breath as I was finally ready to turn the machine on. The entire time, the poor man continued to clutch his chest in pain.

"Give him another nitro," I ordered. With the leads finally attached, I turned on the machine, knowing that at least I knew how to read the tracing once it began to print, but nothing . . . nothing . . . nothing! After all of that, the damn machine—the only one in the hospital!—didn't even work.

With my patient now crying in pain, I started to cry as well. And even though I had repeatedly told myself through that long first twenty-four hours that I would not call my resident for help, no matter what, I broke down and paged him.

It felt like it took forever for him to arrive, as I was afraid I was killing this poor vet who had waited so long for me to assess him. When Dr. R., in his crumpled white coat, finally appeared in the doorway, he unhurriedly strode into the room. He glanced at my patient, then at the EKG machine, and then, to my utter embarrassment, he reached down to plug in the machine, which then proceeded to spurt out a normal EKG. Turning back to me, he muttered, "Give your patient Maalox for his gastric reflux—and you'd better hurry so you're not late for my rounds."

I'd almost made it through my first twenty-four hours without having to ask for help. When I finally broke down and paged my resident, something I swore I wouldn't do, it was because in my fatigue I didn't realize I had failed to plug in the EKG machine! It was bad enough that this was my first day and no one knew my work ethic or ability yet, but I also happened to be a blonde! It was going to be a very long year.

Flash-forward a few months. Now in my last month as a medical intern, I was again assigned to the VA. On my very last night of call, I received a page about a patient with chest pain. By now, I was a pro, knowing just where the single EKG machine was hiding out and just how to attach it to the patient, limbs or no limbs, and yes, I even knew how to plug it in. When I arrived, there he was, that very same double amputee vet, clutching his chest in pain. Recognizing this frequent flier, known for his severe gastric reflux, I calmly ordered some Maalox before performing yet another normal EKG.

I shall always be grateful for the lessons learned during my VA experience. I became a master at getting an IV line established on anyone, even if I had to use the delicate foot veins; learned how to establish a central arterial line in the neck for IV access; performed a thoracentesis, a procedure that drains fluid from the pleural space outside the lung; and inserted a chest tube to reinflate a collapsed lung. Some of these procedures would have been a rare opportunity in any other hospital at my stage of training. I am still appalled by the poor quality of health care services afforded to our veterans who deserve so much more from our health care system. They provide a vital role in our society, as they have been willing to lay down their very lives for our freedom, and in return we thank them by underfunding their health care and other services they are entitled to.

For those who will come after me, it's true: you really can't kill a vet, not even if you're a blonde! God bless them for their bravery on our behalf in the wars and in our hospitals where they play a vital role in training our doctors.

CHAPTER 5

Code of Ethics

The ICU is the heart of the hospital, a place where hope and despair coexist, and patients teeter on the edge of life and death. It lies at the hospital's core, like the hub of a wheel, each spoke leading to a room where individuals fight their private battles in isolation. Those who pass through its doors are easy to recognize—doctors in long white coats over green scrubs, each with a magnetic card granting them access to this sacred, life-and-death space. But who truly decides who lives and who dies here? The patient? The family? The doctors? Or perhaps something greater? I often wonder if the spirits of these patients hover above us, watching, trying to decide if they truly want to return or let go. Strangely, those who survive rarely remember their time here, as though they had been suspended between worlds, caught in a place where time and memory blur.

Each day, I took a deep breath as I slid my access card through the reader, unlocking the doors to the ICU. This was where I typically began and ended my day, attending to the patients with the most severe conditions. The ICU was a labyrinth of individual chambers, each enclosed by heavy glass sliding doors. These rooms circled a central area filled with monitors, charts, and dictation stations. The center bustled with activity—green-suited nurses and white-coated doctors, whispering in hushed tones as they bent over charts and patient diagrams, meticulously reviewing vital signs and laboratory results.

Inside the rooms, patients lay motionless, sedated, with tubes snaking everywhere: into their veins, trachea, and down their throats, delivering the essentials of life. The walls were lined with monitors and machines, creating a constant symphony of hums and beeps. The cacophony paradoxically reassuring, indicating that the unrecognizable form lying motionless on that bed was still alive, at least for now.

As a medical intern, I frequently had patients in the ICU. They often started under my care on the general floor, but when their condition worsened, they were transferred to the critical care specialists. I still rounded on them daily as part of the team, deferring the critical management to the ICU team consisting principally of a critical care pulmonologist, but often also a cardiologist, nutritionist, and any other specialist required by the situation.

Harry was well known to me from multiple hospital admissions that year. In his late fifties and weighing in at over four hundred pounds, he suffered from multiple complications of morbid obesity. He was a very likable man with a hearty laugh and a bighearted personality, often poking fun at himself. He was a favorite of the nurses despite the difficulty managing his challenging size. Already plagued with type 2 diabetes, hypertension, cardiac disease, sleep apnea, and peripheral vascular disease, I didn't think there could possibly be anything else. But this time Harry was admitted with a gastrointestinal obstruction as a surgical complication from a gastric bypass procedure performed to help with weight loss. His obesity complicated his postoperative course, such that he developed peritonitis (severe abdominal infection) from leakage into his abdomen resulting in sepsis and before long the entire cascade of multi-system organ failure leading to shock. Sadly, Harry was now one of the ICU bodies waiting in the antechamber where life or death would ultimately be decided.

Harry's situation became more dire by the day. Too critically ill to survive surgery, he was treated aggressively with antibiotics, cardiac pressors to stabilize his shock, and a myriad of other treatments as one bodily system after another started to fail him. Now requiring mechanical ventilation for his failing lungs and hemodialysis for his failing kidneys, he was barely hanging on. But hang on he did, for 110 days on mechanical ventilation in the ICU, completely comatose. He had been on mechanical ventilation so long that he required a tracheostomy to be attached to his ventilator. There were tubes running into his large neck veins. Nutrients were fed through a gastric tube, along with massive amounts of fluids and cardiac pressor medication trying to maintain an adequate blood pressure. He received ongoing antibiotics as he developed one infection after another. Every tube and arterial line was now infected with antibiotic-resistant strains of bacteria despite meticulous care and changing every three days. These lines were just not meant to remain in a person indefinitely. Down to the last two antibiotics for which he had not developed a resistance, and requiring antifungal agents, Harry was now at enormous risk once the bugs became resistant to these treatments, as they surely would.

Every day, without fail, his wife, Bertha, took three buses to get herself to the hospital, where she sat for hours by his side, talking to him, rosary in hand. A middle-aged Black woman with gray hair tied neatly in a bun, perpetually wearing the same housedress and oxford shoes, she appeared weary as she trudged into the ICU each day to spend her afternoons by Harry's bedside. She worked as a cleaning lady for an office building at night where Harry used to work as a security guard before he became so chronically ill. Not having had any children, Harry was all she had. Harry, no longer sedated, had not awakened at all during this time. Despite the best efforts of the doctors and nurses, he lay there motionless, his wonderful laugh and sense of humor robbed from him, now a shell of his former self. As Harry's condition deteriorated, he suffered multiple cardiac arrests per week, sometimes more than one in a single day. And he developed pulmonary emboli, requiring anticoagulation. The Code team came to know Harry well. Every time I heard a code called in the ICU, I sensed it was probably Harry.

Cardiac arrest is that moment when the heart stands still, signaling that the patient is on the brink of death. It triggers an immediate response from the Code team, a dedicated team of doctors and nurses, hearts all beating in unison as they try to bring the patient back to life. The scene is a chaotic one: people barking orders and, in response, a frenzy of individuals, each with a specific job to do, surrounding the patient, establishing IV lines, obtaining arterial blood gases

. . . then one final order, "Clear!" The charged defibrillator is placed on the exposed chest, delivering an electric shock powerful enough to penetrate the body, commanding the heart to resume beating. They wait . . . no rhythm. "Clear again!" A more intense shock is delivered. Either a rhythm is established, or the patient is declared dead.

Alas, even cardiology had no more to offer Harry. He was not considered to be terminal since he did not have a terminal illness such as cancer, but by now we all knew that he had too much wrong with him to recover from all of this. How many times could we continue to put Harry through the torture of changing his lines and shocking his heart? His heart, lungs, kidneys, and his brain were all so very sick that they were no longer functioning. His body was riddled with bacteria and fungus that was becoming increasingly resistant to every antibiotic and antifungal agent. We were running out of options. Tasked with speaking with his wife as she sat by his bedside talking to him as if he could hear her, I tried day after day to relay the hopelessness of the situation to her and ask that she allow us to arrange for counseling to help her accept the inevitable and make him a DNR. "We can still treat his infections, at least until there are no antibiotics left," I told her. "But please, please reconsider allowing us to apply electric shocks to his heart. He has already coded three times this week."

"No, my Harry is going to come back to me. I just know it."

But Harry was trying to tell us all something different . . .

Every hospital has an ethics committee. Bertha was clearly incapable of comprehending the bleakness of the situation. The time had come to consult with them.

The ICU is not for the terminally ill with no hope of survival as those patients mostly choose the Do Not Resuscitate order, commonly referred to as DNR. This means that no artificial means to prolong life are to be employed, such as cardiac resuscitation or mechanical ventilation. The precious resources in the ICU are meant only for those lives that can still benefit from having everything humanly possible done to keep them alive long enough to heal. The unsophisticated family member sometimes misinterprets our attempts at obtaining a DNR as motivated purely for the financial benefit of the hospital, in an effort to limit resources to the poorly insured or uninsured. In fact, the state of Harry's insurance never even came up in discussion. Not once.

The ICU nurse is the guardian of life in this place, a personification of compassion and care. They are tasked with performing the most intimate of treatments as they turn, bathe, insert lines, administer medication, and fine-tune the machines keeping their patients alive. They are the voice of their patient with the family and the doctors, informing them of how their patient is feeling and what he or she needs, something they somehow just know from caring for them, even though their patient is too sick and sedated to speak for themselves.

The ICU is a different place in the middle of the night, without family members coming and going and just a few nurses and doctors quietly going about their shifts. I was in the ER admitting another patient the night I heard the familiar announcement,"Code Blue ICU. Code Blue ICU." Certain it was Harry, I ran as fast as I could to the ICU expecting to see the Code team working away, but to my surprise they had not yet arrived. His blood pressure was falling rapidly.

"Where is the Code team?" I asked breathlessly.

His nurse, the one who knew Harry better than anyone else, perhaps even better than his wife, replied thoughtfully, "They're on their way."

I was left to wonder if it was Harry who made the choice to leave that day, in the early hours of the morning before his wife returned, when the staffing was low and the nurse who had cared for him all this time was by his side. Did he know that his nurse this time would hesitate to call that code, just long enough to ensure it would be too late, because she knew this was what he wanted, that this was the most compassionate thing to do for her patient? Or was it the Code team that decided to respond slowly to the call so it would be too late by the time they arrived to have to cruelly shock him yet again?

In those quiet moments, as dawn approached, it became clear that some choices are made in the silence of understanding, where compassion outweighs the protocols, and the final act of care is a peaceful release.

CHAPTER 6

Jenny's Story

What really happens in that final moment, when we draw our last breath and slip into Death's embrace? Is it like before we were born—no consciousness, just an empty void? Maybe it's the thought of not existing at all that unsettles us the most. Animals, we believe, have no concept of mortality, yet when their time comes, they stop eating, withdraw to a quiet place, and wait for the end. There are countless stories of animals sensing death in others—like Oscar, the nursing home cat who famously predicted over one hundred deaths, curling up beside patients in their final twenty-four hours. Scientists believed Oscar could smell the biochemical changes of dying cells, choosing to comfort those on the brink. But what did he really sense in those moments? And what waits for us when our time comes?

As a physician, I've witnessed the final moments of life many times. Often, patients pass quietly, their hearts simply stopping as they lie unconscious. But there are rare moments when someone awakens, their face lighting up with a look of awe, as if they've glimpsed something beyond. It's those moments that shape how I view death—not as an end, but as a window opening just as the last door of life closes.

I first met seventeen-year-old Jenny at the beginning of my internship year. She was assigned to my service when she was admitted for breast reconstructive surgery following double mastectomies. She'd received substantial doses of radiation several years earlier as part of her treatment for another cancer, non-Hodgkin's lymphoma that had resulted in the breast cancer, a common complication of chest radiation in females under the age of twenty.

Hodgkin's disease, a cancer of the lymphatic organs, is treated with a combination of chemotherapy, radiation, and often a splenectomy (surgical removal of the spleen), with a 90 percent success rate. The spleen, a unique organ about the size of a fist, sits just below the left rib cage. As part of the lymphatic system, it contains lymphocytes, a form of white blood cells particularly adept at fighting infection caused by encapsulated bacteria such as streptococcus. The spleen also functions as a recycling center for old and damaged red blood cells. We can live without our spleen, as our liver can also recycle cells, but we lose some ability to fight infections from encapsulated bacteria without our spleen.

Jenny's cancer was aggressive, but she'd successfully recovered after being treated with all three approaches—chemotherapy, radiation, and splenectomy. Unfortunately, she then developed bilateral breast cancer as a direct complication of radiation to her chest wall. This beautiful young woman, at the very beginning of her life, had already had to endure so much: two cancers requiring two severe and disfiguring treatments. She had looked Death straight in the eye, and having conquered him twice, she was now yearning to be a normal teenager.

Jenny had little to say to me when I came to take her history and perform a physical examination. She seemed to be very uncomfortable just being back in the hospital, and almost embarrassed to be having reconstructive surgery. I think I was chosen to be her intern over my male colleagues out of sensitivity to her situation. All I could see before me was a sweet teenager who wanted more than anything just to be normal. Her surgery went well, and she was discharged within a few days as expected, a double cancer survivor eager to get on with living the rest of her life as ordinary as anyone else.

If only that could have been the case.

Late one evening in May, I was winding up my long internship year on call. Having just returned to the on-call room after admitting the last ER patient of the evening, I found myself slipping into a delicious sleep when that annoying pager blared at me again. *No . . . please! It's too soon. I'm so tired! I just took the last ER admit and my turn can't be up already!* It was my senior resident. He sounded apologetic: "I'm sorry. I know it's not your turn, but we have an admission down here who was your patient earlier this year, so you should take this. She's being admitted with sepsis."

I hurriedly made my way down to the ER and found Jenny in the trauma bay. She was surrounded by several ER attending physicians and nurses who were working swiftly to establish a large bore IV to infuse antibiotics and pressors to try to stabilize her falling blood pressure. Aside from the occasional barked order, the voices were hushed as they surrounded her bed like a choir of angels wearing green scrubs.

Her veins, having collapsed from shock, were unusable, so they were working frantically to establish a line in the large subclavian vein in her neck.

I was suddenly aware that the evening ER charge nurse, Mary, was standing by my side. I had worked under her tutelage many times this year, and she was a favorite of mine, always ready to help the terrified interns. This night was different. She wasn't here to guide me this time. I noticed that she was crying and clutching a paper chart as another nurse tried to hold her back. Jenny was her only child.

Jenny had complained of a mild sore throat earlier that day. Mary tried calling the doctor and keeping her home that night, aware that infection with an encapsulated organism such as streptococcus could kill her quickly, since she no longer had her spleen. But it was Saturday night and Jenny had plans to attend a graduation party with her friends, so she begged her mother to allow her to go, insisting that her sore throat had gone away. Mary was on night duty in the ER when she got the call from Jenny's friend. Jenny had collapsed at the party and Mary knew immediately what had happened. By the time Jenny arrived at the party, the streptococcus had stealthily entered Jenny's bloodstream without the spleen police to deter it, resulting in rapid duplication and invasion. Her immune system was blindsided. Mary called 911 and had her transported to the ER immediately.

The subclavian line finally established, triple antibiotics and fluids were now rapidly infusing into her bloodstream. As her vital signs began stabilizing, the way was finally cleared for the intern to examine her and write her orders for admission. As I approached, Jenny was still in marked distress, but the look in her eyes signaled that she remembered me. Just as I applied my stethoscope to her chest, she bolted upright with frightful eyes and exclaimed, "I'm going to die!"

At that very moment her heart stopped. I called the Code Blue as the ER team surged around me, their efforts to resuscitate her both frantic and precise. For ninety minutes, the battle against Death raged on. Finally, they stepped back, and the time of death was called. Death, which had failed twice before, returned to claim her spirit at last.

The look in her eyes at the very moment her heart stopped, as she realized she was going to die, is something I shall never forget.

CHAPTER 7

Mother's Day

Denial can be a powerful shield, one that many people instinctively raise when confronted with a life-threatening illness. It allows them to carry on as if nothing has changed, protecting them from the weight of fear and uncertainty—at least for a while. But denial, when left unchecked, can become a dangerous force. For some, it builds walls so high that even the thought of seeking medical help feels impossible. The consequences of that refusal, however, are all too real, turning a momentary escape into a path toward disaster. This is where she found herself—a young woman, vibrant and full of life, clinging to hope and denial as the signs of breast cancer crept ever closer.

It was my last admission of the day before I got to go home to my family. It had been a long thirty-six hours on call. *Will this internship year ever end?* This admission was a young African American woman aged about thirty-five. Given her young age, I anticipated that it should be straightforward. That wasn't to be the case.

As I approached her room, I was greeted by an older woman—her mother, I presumed—and three adorable young children, two girls and a boy.

"Please help Keisha," the woman said. "I have been begging her to see a doctor for months, but she just wouldn't listen to me. She has been working so hard holding down two jobs and trying to complete her college degree in teaching that I think she just exhausted herself. I have been helping raise her three children since her husband left them three years ago. A few weeks ago, she went to bed, and I've been unable to get her up since. She barely eats or drinks and has lost so much weight. Her fiancé has been unable to get through to her either."

Keisha asked for complete privacy when I went in to examine her, hoping to hide the seriousness of her condition from her concerned family. Despite her emaciated appearance, I could see right away how beautiful she was, with large almond-shaped brown eyes and cascading dark curls framing her ebony face. Sitting up, she tearfully confessed that she had noticed a lump in her breast about six months ago. At first, she thought it had something to do with hormonal issues, but as the months wore on and it doubled in size, she feared it might be cancerous. Too terrified to find out for sure, she made up all kinds of excuses not to seek help. After all, she was too young for cancer, right? She was aware of a family history of breast cancer, but they had all been much older than her. "I have been working day and night to support my family while trying to get an education," she said. "I had no time to see a doctor, and I didn't know how I would even pay for it, since I have no health insurance. I need to be OK. I have three children, and I am about to marry the most wonderful man in the world."

I think if her mother had not forced her to come to the ER, Keisha would still be at home in bed hoping this would all just go away. She had not felt well for a considerable time, with constant nausea, fatigue, and back pain.

My examination revealed a lump in her left breast the size of a golf ball, in addition to numerous palpable lymph nodes under her armpit, a sure sign of metastasis. I ordered a mammogram, chest and abdominal CT, bone scan and blood work, in addition to requesting a consult with oncology.

The news was devastating: triple-negative breast cancer (TNBC), meaning the cancer cells didn't have estrogen or progesterone receptors, and also didn't make any or too much of the protein HER2, resulting in it being extremely difficult to treat. It differs from other types of breast cancer in that it tends to grow and spread faster, has fewer treatment options, and tends to have a worse prognosis. Keisha's cancer was already widely metastatic to bone, liver, and brain. This type of cancer is extremely aggressive, more likely to strike a premenopausal Black woman who is positive for the BRCA1 gene (a gene discovered for a familial link to breast cancer). But this was 1990, four years before this gene was discovered and six years before commercial testing would be available. Today I'm left wondering if that knowledge would even have mattered in this particular situation, given how strong her denial was, coupled with her lack of medical insurance. Usually with this type of cancer there is up to a 30 percent chance of survival in five years with aggressive chemotherapy, but not in Keisha's case because she presented so late. Of course, they would try everything in hope of some palliation and time, even just a few more months.

As the weeks progressed, Keisha submitted to the ravages of radiation and chemotherapy—her final weapons in the fight—which exacted their own toll on her fragile body. I would see them often—three young children and her mother, accompanied by a handsome young man. They were usually waiting outside the cancer infusion suite, or radiation department, collectively and bravely fighting in their minds this sinister invader that threatened to take their most cherished loved one. Whenever I saw them, I would stop and ask how she was doing. They seemed hopeful at first, but later more resigned to the inevitable.

I was on call for internal medicine again that Mother's Day when I was paged to the oncology ward for a patient with headache and nausea. When I entered her room, filled with the most beautiful, fragrant flowers, Keisha remembered me right away as the intern who had admitted her a few months ago. Her black curls now replaced by a scarf around her balding head, she was still strikingly beautiful. Surrounding her bed were her three small children and her handsome fiancé. I was initially struck by an unusual serenity about them as they waited for me to perform my assessment and prescribe something to ease her discomfort.

Wishing her a Happy Mother's Day as I left, I turned and asked if there was anything else that I could do for her. Looking hauntingly into my eyes, she replied, "Yes, a cure, please. I just so want a little more time."

I left feeling utterly powerless to help her, to help them. Denial is a powerful defense mechanism, often used to shield us from painful emotions. But for this mother, who had fought valiantly to gain an education to support her young family and had finally met the man of her dreams, denial became a cruel adversary. Her refusal to acknowledge that something was terribly wrong robbed her of the life she had longed for just as it was beginning.

PART THREE

Neurology Residency

Stepping Into My Own

CHAPTER 8

Every Day Aboveground Is a Good Day

The day began like any other, with my alarm ringing at 5:00 a.m. I hoped my early start would give me a few precious moments with my young son before heading to the hospital. Summer traffic into Philadelphia was often lighter, with schools out and families on vacation, so I could make the drive in under forty minutes. Those early morning drives had become a rare moment of calm, a quiet space to gather my thoughts before the chaos of the day began. After surviving the grueling first year of my medical internship, I had finally stepped into my neurology residency—a new chapter, full of challenges, but one I was eager to face.

The drive to the hospital that morning was uneventful. Reaching the parking garage with five minutes to spare, I drove to the very top of the garage, giving me farther to walk, probably my only exercise for the day. As I exited the garage, I could see the stately hospital just across the street, beckoning me in, surrounded by historic Philadelphia's cobblestone streets and their streetlamps in the style of colonial-era gas lights. The rising sun was almost blinding, the light filtered by the trees swaying in the crisp early morning air. At that moment, an image I had never seen before suddenly sprang vividly into my mind's eye, unbidden: that of my own mother, her face pale and lifeless, in an ICU, on a ventilator with tubes and IV bags everywhere. The head of her bed was tilted in a downward Trendelenburg position, as in cases of hypotension and shock. Somehow, I knew at that moment that she was dying, and the phrase *atrial fibrillation*, an irregular heart rhythm associated with high risk of stroke, sounded in my mind. The image was so vivid it jolted me, but just as quickly as it came, it was gone. I shook it off. My mother had been perfectly fine the week before, laughing and full of life. There was no reason to think something was wrong. Seconds later, as if it never happened, I found myself crossing the street, about to walk through the hospital's heavy glass entry doors.

I felt certain the image was just due to an overactive imagination. After all, I had just started my neurology residency. Having spent the last few weeks managing stroke cases in the ICU, atrial fibrillation requiring anticoagulation (blood thinners) had been a common complication. Atrial fibrillation is an irregular, often rapid heart rhythm that can potentiate the formation of blood clots in the heart, which can then embolize to the brain. The stroke risk associated with this is often treated with blood thinners to prevent the blood clots from forming. I was struggling to learn as much as I could as fast as I could to care for these patients, having been a neurology resident for only three weeks.

Little did I know that brief, haunting glimpse would become a chilling reality.

The days and nights went by quickly, filled with learning as I worked under the guidance of senior residents and neurology attending staff. The on-call experience was particularly challenging, as it was just me responsible for these very sick patients during the middle of the night. The mere thought of being the "only neurologist" in the hospital at night, having only been practicing neurology for a few weeks, was something I dared not even contemplate when I received that call to come to the ER for a "consult." Just a few short weeks into my first year, there I was, being called to the neuro-ICU with a patient crashing with elevated intracranial pressure or a cerebral bleed, and me with only my *Little Black Book of Neurology* pocket guide to help me. There were times during these hectic days when the patients all seemed to blur together: cerebral hemorrhages, status epileptics, unstable elevated intracranial pressure, strokes, strokes, and more strokes, all lying there attached to various forms of life support. My neurology backup attending was at home, snugly tucked into his bed, and I would do almost anything to avoid waking him.

The neuro-ICU was filled with people, ordinary people—someone's parent, someone's child, someone's spouse, just like you and me—but I rarely got to meet them before their event. So, by the time I arrived on the scene, they were already sedated and had tubes and machines everywhere. I only got to humanize them later through the stories their loved ones would tell me.

I spent my time following their numbers: vital signs, input and output, ICP (intracranial pressure), ventilator readings, etc. I had been trained to be objective, to distance myself from their pain so that I could be decisive and treat their disease, and so that after a failed code, I would be able to walk down the hall and start all over again, objectively assessing the next patient, the next stroke, the next seizure. I felt like I was beginning to master that ability, but I was still struggling to master the art and science of the neurology itself.

But what exactly was that ICU vision of my mother I had experienced that morning? A hallucination, a daydream, a thought? Just my overactive imagination with intruding images derived from a compilation of the many images I saw daily, especially rounding in the ICU? It seemed like a glimpse of a dream, but I was most definitely awake. I had seen so many similar images over the past several weeks in real life, one patient after another in the ICU, on ventilators, sedated and attached to tubes and IVs, each critically ill and many with a prognosis poorer than the last.

I had seen my mother just a few weeks earlier, during the transition week between my internship and residency. I had taken a brief trip to the beach at Cape May with my husband and son, and she drove down for an afternoon to spend time with us. She was fine. She *is* fine. Time spent with family during medical training was precious, and we all enjoyed a wonderful week together. Putting the image out of my mind, I entered the main lobby and made my way to meet my senior resident to get my assignments for the day.

It was always a good day when I wasn't on call, or when I had completed call for the day and was free to enjoy an evening with my family. Forgetting all about my disturbing premonition that morning, I was talking about the events of the day with my husband when the phone rang. It was my sister, Judy.

"Mom was just admitted to the hospital with an upper respiratory infection and difficulty breathing," she said. "They had to put her on a ventilator, but they expect she'll be better in a few days on antibiotics. So no need to worry."

Suddenly, my apparition came flooding back to me now. "Did they say anything about atrial fibrillation?" I asked.

"Oh yeah, forgot to tell you that part."

My mother's hospital was eighty miles away, and after calling the hospital, I was assured she was stable for the evening. I was on call the following night, which meant I would be working at my hospital for the next thirty-six hours. Thoughts of my mother intruded into my concentration throughout that long shift. *There's no reason for alarm; she has not had a heart attack or stroke. She's only on a ventilator because of an upper respiratory infection that probably triggered the irregular cardiac rhythm, and she should already be improving by the time I get there.*

When my thirty-six-hour shift finally ended, exhausted, I raced to the community hospital at the Jersey Shore where my mother had been admitted. This wasn't the kind of ICU I was accustomed to at my large academic institution. There were only four beds, separated by white curtains instead of glass doors, but the combination of the sounds of the ventilators and heart monitors and the antiseptic smell were unmistakable. I was relieved to find my mother awake and sitting up in bed, her crystal blue eyes, the same eyes that could always see right into my soul, trying to console me. Her wonderful dry sense of humor was on full display on the sign she had the nurses put above her bed: EVERY DAY ABOVEGROUND IS A GOOD DAY.

Despite adequate treatment of her lung infection with antibiotics, the doctors were having a difficult time getting her off the ventilator because of her chronic pulmonary disease from her history of smoking. They had also started her on IV heparin to thin her blood, as a preventive measure for stroke, because of her new atrial fibrillation that had the potential to form blood clots in the heart. Communicating to me through handwriting, she related that both a pulmonologist and cardiologist were on her case, and she complained to me about abdominal pain. I asked if she had let her doctors know; she confirmed that she had. Before leaving, I inquired about this with the nurse at the desk, but the doctors had left for the day. She kindly placed a call to the pulmonologist so I could speak with him. He said he was aware of her complaint and thought her pain was just from coughing from the ventilator. But he promised to call a surgical consult that evening to have it checked out further. Tired but reassured, I drove home and crawled into bed, only to have to get up three hours later to go back to work. It was going to be a very long day.

My mother looked good and was in good spirits. She was going to be all right. I had twenty inpatients to round on, some in the ICU, and the floor consults were already starting to come in. My pager was going off constantly, each call taking me in a different direction, to a different crisis, or just to talk to a worried family member again. It felt good to be busy and productive, so I had less time to be distracted into worry.

Then came the call.

It was 10:00 a.m. and my mother had been found unconscious that morning, having bled out nine pints of blood into her abdominal cavity overnight. Her heparin level had been critically elevated the day before and never adjusted. Her abdominal pain had been from the bleeding, which was completely missed by her pulmonary attending and the surgical consultant who came in the evening before. He had just given her morphine without ordering any further diagnostic studies, failing to notice her critical heparin level, which had been missed earlier. The sedation from the morphine further contributed to the delay in discovering the bleeding.

I knew at that paralyzing moment that I would never speak with my mother again, that she had already crossed a threshold that meant she was incapable of *meaningful neurological recovery*. How many times I have used that phrase over the years, and how many times I have witnessed the look of terror and loss in the faces of my patients' loved ones.

I raced to her hospital, begging and pleading with God to somehow spare her. It was not yet her time. She would never get to see her grandson grow up, and Rory would not have the delight in knowing her. And what about all her other grandchildren and family memories yet to be created? She had worked so hard her whole life, raising four children all by herself. This was supposed to be her time to enjoy her family.

And what about me? I wasn't ready to lose her.

When I arrived, there was the unmistakable foul smell of melena—black tarry stool from gastrointestinal bleeding—filling the air. Her bed was in Trendelenburg and there were numerous IVs infusing blood pressure pressors and blood.

She was dying, and the image was exactly as I had seen it that morning three days before, in the rising sunlight. She never regained consciousness. She lost oxygen to her brain and her kidneys from critical loss of blood.

For the next five days, my mother remained in that state, occupying the corner room in the ICU, surrounded by starched white curtains, the light from a small window across the hall illuminating the pallor of her skin. Her four children kept constant vigil by her bedside, as we didn't want her to die alone. Late one evening, I had just stolen a few hours at home to see my little boy when the call came. It was Judy.

"Mom's heart just stopped, and she's been pronounced dead. The doctors want to know if you would like them to keep her body in the ICU to say goodbye, before sending her to the morgue?"

"Yes," I said. "I'm leaving right now. Please keep her where she is."

I raced back to the hospital, and upon arrival, I was surprised to see my three siblings still surrounding her bed. As I approached, I could hear the *whoosh . . . click . . . whoosh . . . click* of the respirator and the undeniable heartbeat from the monitor.

"I don't understand," I said. "I thought you said her heart had stopped?"

"The most unbelievable thing happened when I returned from making the call to you," Judy replied. "Mom's heart just started up again. It's slow, but she's not dead!"

For the next few hours, we all stood around her bed recalling wonderful stories, funny stories, mostly about her sense of humor. As the sun started to rise, casting a beautiful orange and pink glow on her bed, her heart stopped for a second time—this time the last. I have since imagined my mother's spirit hovering over her bed that night, listening to my call announcing I was on my way, and waiting for me so she could spend a few more moments alone with all her children together. She would have loved that.

My mother died on Thursday. We had a funeral for her Saturday. I returned to my resident duties on call at the hospital on Sunday. I was unable to make it through that first day, collapsing in tears in the restroom. I was sobbing so hard that I couldn't even manage to answer my pager. A very kind neurosurgical resident, who had been looking for me to sign out his patients, found me there and sent me home for the rest of the day. For months afterward, I struggled to get through my days. I dutifully showed up, took all my overnight calls, but my heart was not in it anymore.

Losing my mother so early in my career shattered my sense of objectivity as a physician. Her death, which could have been prevented, left me questioning everything—my choice of specialty, my future in medicine. For nearly a year, I struggled, tempted to walk away from neurology, and from medicine entirely. But slowly, as I worked through my grief, I found my way back. In time, I regained the ability to care for patients with clarity, though my compassion had deepened in ways I hadn't expected.

If there's one thing I took from that painful experience, it's the power of intuition. It has become my greatest ally in medicine, a gift I've learned to trust and respect. And through it all, I carry my mother's memory with me—not just for her strength and courage, but for her ability to laugh, even in the darkest moments. It's that spirit I hold on to, and that keeps me going.

CHAPTER 9

Misdirected Guilt

At first glance, this hospital seems inviting, almost serene. The soaring glass atrium, draped with lush greenery, the deep crimson carpets, and the carefully chosen art on the walls—all of it crafted to comfort the visitor. But beneath the polished facade, this place remains what it truly is: a place where people die. They die quietly in our beds, under the harsh lights of our tables, even in the small, silent cribs. No matter how we dress it up, death lingers here, always just out of sight.

It was Easter Sunday 1991. I was the first-year neurology resident on call when I was summoned to the maternity ward to evaluate a severe headache in Marcia, a forty-year-old woman who had recently given birth to her fifth child. Even though she had a long-standing history of migraine, this headache was lasting too long, despite conventional treatment, prompting sufficient concern to warrant a specialty consult.

As I made my way to see Marcia, I reviewed the differential diagnosis of postpartum headache in my head: migraine, preeclampsia and eclampsia, meningitis, acute ischemic stroke, intracerebral and subarachnoid hemorrhage, cerebral venous sinus thrombosis, posterior reversible encephalopathy syndrome, and reversible cerebral vasoconstriction syndrome. Nine months into my neurology residency, I had more book knowledge than clinical experience, and I didn't want to miss anything. Alone on weekend call, it was always comforting to know there was a full-fledged attending neurologist backing me up at home, just a phone call away.

Entering the doorway to Marcia's room, I felt like a spy as I observed this beautiful woman sitting up in bed with her newborn baby boy clutched to her breast. The sunlight streaming through the window illuminated her silhouette, and right away I sensed something off about her. Watching her with her child reminded me of how I longed to be at home with my three-year-old son on this Easter morning. Standing in my blue scrubs, long white coat, and pearl necklace, I broke the silence, announcing that I was here to perform a neurology consultation for her persistent headache. She seemed barely conscious of my presence as she stared blankly past me, the look in her eyes screaming that her distress was more than mere physical pain. The mood in her room that day was heavy with a despair that hung like a cloak over a corpse.

Assessing her carefully for any evidence of meningitis or increased intracranial pressure, I noted that her neck was supple, and her eye exam showed no evidence of papilledema (swelling of the optic nerve). Finding nothing out of the ordinary, I was still concerned that something more serious than a prolonged migraine was going on. I decided to order a brain MRI and venogram to assess for a structural or inflammatory process such as a tumor, infection, swelling, or a blockage in the sinuses that drained the blood from the brain as the list of postpartum complications was vast. Feeling there was enough concern for urgency, I called the MRI technician in to do the study—not an easy feat on Easter Sunday.

"Can't this wait until tomorrow?" he said.

Still unsure of myself at this stage of my career, I shakily answered, "No. I need this now."

Two hours later, studies completed, I reviewed the MRI with the radiologist. He confirmed that Marcia had an evolving cerebral venous sinus thrombosis: a blood clot had formed in the sinuses of the brain, from which the blood drains back down into the vascular system, creating a blockage. One of the most common causes of this condition is a state in which the blood coagulates more than normal, typically associated with elevated levels of estrogen, either from hormone replacement therapy or naturally during the postpartum period. Symptoms begin with headache, but as the condition progresses and pressure from the vascular blockage builds, red blood cells sneak into places where they don't belong. They get into the brain tissue, which can quickly result in a hemorrhagic stroke. Treatment for this thrombotic condition often consists of anticoagulation to thin the blood before more serious complications arise, but it's a controversial double-edged sword, given that anticoagulation medication also increases the risks of bleeding.

Realizing I lacked the experience to deal with this complicated life-threatening condition at this stage of my training, I called in my attending neurologist to review the case with me. As it turned out, he knew the patient and her family personally—he had managed Marcia's migraines for years—so his decision-making was complicated, possibly compromised. Worried about the risk of a hemorrhagic stroke and hoping the headache was more migraine related, he decided against anticoagulation initially, in favor of treating her migraine more aggressively.

Marcia's condition deteriorated within hours. She developed paralysis on one side of her body from bleeding into the brain—a hemorrhagic stroke. Anticoagulation was now started very slowly in the hope that it would prevent further clotting without further complicating the bleeding. Had we chosen anticoagulation initially she probably would have bled anyway, but we never would have known if it was a complication of the treatment or the natural course of the thrombosis.

Marcia was transferred out of the maternity unit where she was close to her baby to the neuro-ICU, a windowless cave with sterile white walls and sleek machines, where her condition could be monitored more intensely. Marcia seemed to stabilize neurologically. On my daily rounds, however, she remained hemiplegic, with complete paralysis on her entire left side, and she refused to even look at her baby boy when they brought him to her.

One day, as I was leaving her room, a good-looking middle-aged man in a white button-down shirt approached me, desperate to talk.

"You must be my wife's neurologist," he said. "I have been waiting to speak with you. Do you believe in karma, Doctor? Marcia is so depressed; she can't bear to look at me or our baby. I can't help but feel this is all my fault—our fault. You see, we planned this pregnancy out of sheer anguish, despite Marcia's age and the fact we already had four children at home, because we chose to abort another pregnancy a year earlier, a decision we deeply regretted. It was an unplanned pregnancy, and I was struggling to get that promotion I had been working for and we just didn't know how we would ever manage a fifth child under such stress. I was the driving factor, I'm afraid, and Marcia never got over it.

"So here we are, and look at what has happened!"

Her husband's guilt and fear for the future was heavy in the air.

"I don't think anything of the sort," I replied. "Marcia's prior abortion has nothing at all to do with what is happening to her now, and it likely would have happened with the prior baby had she gone through with that pregnancy. Terminating a pregnancy is a very personal decision and you were trying to do the best you could for your wife and your other four children at the time. Your wife is suffering from a postpartum depression, and this can be treated. The type of stroke she suffered, a result of a venous rather than arterial hemorrhage, also carries a good prognosis for recovery over time."

As hard as I tried to reassure him, however, I sensed that my words were falling on deaf ears.

That night I was on overnight call again when I was paged urgently to the neuro-ICU. Marcia was deteriorating rapidly. She was now unconscious, with her head arched backward. Her arms and legs were outstretched stiffly, her toes pointed downward, a sign of decerebrate posturing, indicating massive brain swelling, usually a terminal event. I called in the neuro-surgical resident and together we worked on her throughout the night in a desperate effort to save her life. The anesthesia team was quick to arrive to intubate her so that we could set up the mechanical ventilator to artificially hyperventilate her, a temporary means of acutely reducing cerebral edema by blowing off carbon dioxide. Intravenous mannitol was begun as another temporizing method to reduce the cerebral blood volume. As Marcia continued to deteriorate before our eyes despite all of this, the neurosurgical resident performed a bedside ventriculostomy, to further reduce the intracranial pressure. I assisted as he positioned a small drill just behind her hairline that whirred for just seconds until a sufficient burr hole was established large enough to pass a catheter through the brain and into the frontal horn of the lateral ventricle. The distal end of the catheter was then deftly tunneled through the scalp to an outside bedside collection system in the hope that this would be sufficient to reduce her pressure further by draining some cerebral spinal fluid. As the collective sounds of the ICU—the click of the ventilator and steady beep of the heart monitors—echoed in the background, we remained silent as we worked, knowing the outcome was likely to be very grim.

By early morning, Marcia was brain-dead despite our best efforts. The weary neurosurgical resident looked right into my eyes as he muttered, "Losing this young mother is my worst nightmare."

It was a nightmare, indeed.

There was a time when this couple didn't know how they would ever manage an unexpected fifth child together. Now, this widowed young father would need to care for and support a new infant in addition to four other children. His baby boy, who had yet to be given a name, would never know his mom. I will always recall that conversation we had about this tragic event representing some sort of karma for the decision he and Marcia had both made to abort another child under severe stress. I wish I could have gotten through to him that day that life just happens, as cruel as it is, that they did the best they could with the knowledge they had at the time. If there is such a thing as karma and an afterlife, I have hope that Marcia is reunited with her unborn baby.

Everyone has their own set of values and beliefs, shaped by their culture, religion, and scientific knowledge. It's difficult to argue either in favor of or against abortion without considering all sides of the debate. There are situations in which another child poses significant financial hardship, then there are situations such as incest, rape, or an underage mother, and others in which there is advance knowledge of a devastating fetal abnormality. No matter where one stands on this complex issue, most women I've come across in my career did not make this decision lightly. In fact, many of them have gone on to agonize over it for years afterward. This speaks to the gravity of the decision, and the deep moral consideration that goes into it. It also speaks to the strength and resilience of these women, who are often faced with difficult circumstances and still make the brave decision to take control of their lives and futures.

As do we all when it comes to matters of the heart and soul.

CHAPTER 10

Right Brain, Left Brain

Did you ever wonder what would happen if your left brain could not communicate with your right brain? Well, buckle up for a mind-bending adventure! Your left brain is the one you wouldn't want to leave home without; it is the cautious, logical guardian, the part that monitors your behavior. Your right brain is the mischievous troublemaker, the one that, if left to its own devices, will get you into trouble every time. It's like having a responsible chaperone and an impulsive party animal in your head!

People experience this tumultuous conflict all the time when they get intoxicated with alcohol, essentially putting their left brain to sleep so their right brain gets to play unsupervised. With the left brain's judgment on vacation, the right brain goes wild, leading to reckless escapades, consequences be damned! Stripped of its discerning judgment, the left brain submits to the right brain's impulsive urges, surrendering control to reckless abandon. As the right brain parties unsupervised, consequences blur, drowned in the sea of ephemeral pleasure, and the moral compass falters, steering astray from the path of reason.

But wait, there's more! In the realm of neurosurgery, there's a daring procedure—the corpus callosotomy—for severe epilepsy cases. They literally cut the connection between the left and right brain hemispheres—the corpus callosum. Picture a high-stakes surgical procedure where the goal is to partially sever the connection, preventing seizures from spreading from one hemisphere to the other and thus preventing the unconsciousness that occurs with a generalized seizure.

Epilepsy, a brain disorder fueled by electrical chaos, disrupts the orderly arrangement of brain waves. This occurs when a spark ignites, resulting in an electrifying storm. It can be caused by scarring, or other reasons we don't fully understand. It's like the brain's equivalent of a thunderstorm, complete with a range of sensory and motor fireworks, from familiar smells and strange sensations to convulsive movements.

Many people with epilepsy lead perfectly normal lives, except for the occasional circuitry disruptions. Thankfully, modern medicine has given us clever tools to manage seizures with selective medications, taming the unruly electrical currents and keeping the sparks from spreading like wildfire.

But some individuals have epilepsy so severe that it feels like an endless fireworks show, with up to a hundred seizures a day. These brave souls require multiple medications just to function. Their lives become a maze of restrictions, where driving, cycling, and even taking a bath alone are deemed too risky.

In desperate cases, where other treatments fail, a corpus callosotomy procedure steps in to save the day. By strategically disrupting the path between hemispheres, it prevents the spark from crossing over and overwhelming consciousness. However, this adventure comes with a catch: the corpus callosum must be only partially severed, preserving vital communication channels between the left and right brain.

During my neurology training, I met an extraordinary woman named Sally, who had undergone a corpus callosotomy. She delighted in beguiling the neurology residents with her incredible story, holding us captive with her humor and antics. It was like watching a live comedy show, with Sally as the star performer.

As we entered her hospital room, we couldn't help but notice Sally sitting up, her hands playfully hidden beneath her. She opened up about her challenging life, telling us how her seizures had robbed her of a normal childhood and confined her to a helmet-wearing existence to prevent recurrent head injuries from frequent loss of consciousness.

Countless medications had left her zoned out, barely functioning. Her first corpus callosotomy procedure failed. But thanks to her second, more successful callosotomy, her seizures were under control, granting her the freedom to venture beyond the walls of her home at last.

However, as Sally shared her tale, something extraordinary unfolded before our eyes. Her mischievous right brain decided to rebel, and her left hand, under its control, attempted repeatedly to light a cigarette, while her right hand, desperately trying to restore order, extinguished the flames.

The battle between her hands became a hilarious spectacle, like a slapstick comedy routine.

And that wasn't all! When a good-looking young resident entered the room, Sally's left hand mischievously unbuttoned her shirt, while her right hand valiantly fought to restore decency. Sally maintained a calm demeanor, oblivious to the absurdity of her situation. It was as if her brain had split into two distinct personalities, each vying for control. The right brain reveled in the chaos, while the left brain carried on with the storytelling, acting as if nothing was out of the ordinary.

Interestingly, we discovered that Sally's left-brain speech center remained intact, so she sounded perfectly normal while sharing her experiences. She recounted funny anecdotes, like her struggles with riding a bicycle after the surgery, where her left and right sides competed for control, resulting in a comical balancing act as one side wanted to go one way while the other side wanted to go another.

Despite the life-disrupting nature of her peculiar condition, Sally approached it with amusement, at least from her right brain's perspective. Having been confined to her home for most of her life, she remained grateful for the surgery, as it had granted her the freedom to venture into the world without fear of sudden seizures. It was a bittersweet victory, though, showcasing the complexities of the human brain and the triumph of overcoming immense challenges.

And it taught all of us just how different the two sides of our brain are with respect to behavior.

CHAPTER 11

Righteous Intolerance

There was something sinister lurking in the bloodstream of those afflicted with this mysterious condition—a microscopic invader, the likes of which we had never seen; invisible to the human eye, yet its impact was devastating. This thing, a ruthless and relentless killer, infiltrated the body with a quiet and deadly efficiency. Like a merciless hunter, it sought out its prey—the immune T cells—and devoured them, leaving destruction and chaos in its wake. Without a functioning immune system, its victims would suffer a slow and painful decline, their strength and vigor sapped, as the killer took its toll. Its powerful grip spread insidiously through the bloodstream to every organ, every cell, leaving behind a wake of suffering, despair, and death.

This was 1992, the height of the AIDS epidemic. There was no cure. All we could do was try to treat the infections and the cancers that afflicted those without a functioning immune system and give AZT, an antiviral drug that was about as effective as trying to nail Jell-O to the wall.

With something so deadly on the loose, something that was claiming the lives of hundreds of thousands of young people on the threshold of their lives, one would expect the public to be up in arms at this injustice to humanity and threat to our children. Money and resources should have been pouring in from everywhere to battle this horrible malady. Only that was not happening.

But why?

The Reagan administration's response to the epidemic was marked by inaction and denial. Despite mounting evidence of the disease's severity, President Reagan and his advisors delayed any meaningful action for years. During this time, federal funding for medical research was slashed, severely hindering the efforts of scientists and medical professionals to address the growing crisis. Not only did they fail to acknowledge the seriousness of the epidemic, but they also implemented policies that further obstructed access to care for those afflicted, leading to the unnecessary loss of countless lives. To the administration, this was not a public health crisis—it was a "social disease," one primarily affecting a group they deemed morally deviant: homosexual men. The addition of IV drug users to the list of the afflicted only deepened the administration's discriminatory stance. Measures that could have curbed the spread of the disease, such as needle exchange programs, were actively opposed, leaving those most at risk even more vulnerable.

This was a time when homosexuality was still widely seen as a moral failing or even a mental illness in the United States.

The LGBTQ community faced rampant prejudice and discrimination, with same-sex marriage not recognized in most states. The military's 1993 "Don't Ask, Don't Tell" policy, which barred openly gay and lesbian individuals from serving, further institutionalized this discrimination, reflecting the broader societal rejection of LGBTQ rights.

The stigmatization caused many homosexuals of the day to remain in hiding, their true nature invisible to family and friends. The epidemic, however, forced many of them to finally come out in the open. Just at this time, when they were so sick that they needed their family the most, they were often deserted out of fear and disgust. So that all they had were their friends in the gay community and the caring nurses and doctors who volunteered to treat them.

The concept of adhering to universal precautions against blood-borne diseases was born out of this epidemic, in an effort to shield the patient seeking care from being shunned due to the widespread misinformation and prejudice, which was everywhere—in our government, in our communities, and in our hospitals. Everywhere. And those so afflicted were often easy to spot, with the characteristic reddish-brown skin lesions from Kaposi sarcoma (a rare skin cancer common in the immune compromised), the persistent cough suggesting pneumocystis pneumonia, enormous weight loss, and the characteristic slowing of cognitive processes, as the invader took up residence in the brain.

Fear gripped the population most at risk, as they struggled to get through the days, trying to keep busy and productive while supporting their community. No one knew when their time might come. Would today be the day they would develop a cough, find a suspicious reddish-brown skin lesion, succumb to debilitating fatigue, develop night sweats? How long could they hide their symptoms from their coworkers, their family? They endured endless hours of torture waiting for their blood test results. Every time the phone rang, their heart leaped in anguish as they picked up the receiver with sweaty palms, anticipating that this might be their certain death sentence.

As a doctor in training, I understood that I couldn't become infected by merely touching a patient. But even so, I felt despair at caring for them. There was so little I could do to ease their suffering, especially the emotional toll that this disease took. It was all just so unfair! I came to understand just a small inkling of their anguish after the occasional, unintentional self-inflicted "needle stick," usually the result of sheer exhaustion on my part. Such an incident generated endless paperwork, mandatory initial testing, and testing yet again six months later. All I could do was to keep as busy as I could while I hoped and prayed that I would be OK. Otherwise, I would spend my time obsessing over the latest statistics on my risk, according to the viral load of my patient.

As a neurology resident, I treated every aspect of HIV that affected the nervous system. The virus infiltrates like a slow-moving shadow, creeping through the body and weaving itself into the very fabric of the brain and spinal cord. At first the symptoms are almost imperceptible. In its final stages, HIV can cause a cascade of neurological disorders, from debilitating pain that radiates through the nerves like electric fire, to dementia that clouds the mind in forgetfulness, leaving behind a trail of confusion, loss, and disconnection from the world.

Hospitals were filled with patients so afflicted, often marginalized in special wards or isolation rooms wearing the stigma of contagion. My own teaching hospital had a floor dedicated to those so afflicted. As I peered in from the doorway, my patient's back was toward me, as he stared at the opposite wall in despair. I called his name; there was no answer. I approached closer and called again. Slowly, his face turned toward me, his vacant eyes sunken in his wasted skull. I could only imagine how this young man must have looked just a short time ago, before the invader ravaged his immune system, stealing all his strength and hope.

"Is there anyone I can call for you? Anything you need?" His vacant eyes stared back, but he declined to answer, turning away from me again. There were no flowers, no cards, no sign of friends or family in the room.

Teaching rounds were a glorious escape from the suffering on the wards. As neurology residents, we had the rare opportunity to intellectualize the situation in theoretical discussions with our superiors and peers as we would go from one sad patient case to another. The time I spent in these intellectual rounds reaffirmed for me the specialty I had chosen.

One such morning in late June, I had the enormous privilege of rounding with one of our program's senior residents. George was an exceptional teacher, as brilliant as he was flamboyant, kind beyond measure, and generous with his knowledge. A recipient of a coveted academic scholarship from the most distinguished of institutions and scion of a respected family, he was bound for one of the most prestigious neuro-ophthalmology fellowships in just one month, following completion of his residency.

George was tall, with an athletic build, reminding me that he was a competitive tennis player. He was good-looking with a boyish grin, brown curls, and bright blue eyes. As he stood in front of the group wearing his signature bow tie, I caught a glimpse of the characteristic reddish-brown skin lesion, extending from just below his starched white clinic jacket onto his wrist. I said nothing. At that moment, I knew George would never realize his hard-earned dreams. He would never make it to his fellowship. In the coming months, he was destined to be shunned by his own academic home. He would be given a job at a competing institution, where, as an attending neurologist, he would spend the next eighteen months of his life teaching, until he died alone in the hospital with no one by his side except his dear friends, his own family having disowned him in his time of greatest need.

The unjust ostracization of the AIDS patients in those early years of the epidemic was, in many ways, so much worse than the virus itself. George, as I recall him, was so much more than a brilliant neurologist with a promising career. He was a bright light in a dark time, an example of courage. In the face of an agonizing fate, his spirit found the strength to show that there was still beauty and hope in this world by spending his remaining time here helping and teaching others.

CHAPTER 12

Grief Waits

Grief is like a shadow that lingers at the edge of the soul, an unwelcome guest that arrives without warning and refuses to leave. It is a heavy cloak, draped over the heart, pressing down with a weight that seems at times unbearable. In the quiet hours of the night, it whispers like the wind, reminding one of loss and absence.

One of the things I loved about the practice of neurology was the challenge of diagnosis. Neurologists are often referred those patients with a myriad of complex symptoms no other specialty could quite sort out. We frequently became the last resort, so to speak. I often thought that was because, by the nature of our chosen field, we were somehow more thoughtful and contemplative. My best teachers in the field taught me that the answer was most often found in the history. *Listen to your patient carefully. Never rush them, for they will often give you the answer you are seeking before you even examine them or perform any tests. If you have listened well, your examination and testing merely confirm the diagnosis for you.* I took this to heart as their words flooded back to me in some of my most difficult cases.

Martha was a sixty-year-old woman referred to our neurology residency specialty clinic for evaluation and management of advanced dementia such that she had become unable to function on her own. She was accompanied by an entourage of concerned family: her husband, son, and two daughters. Her husband, a balding older man with kind eyes, began by saying, "We've been everywhere, had every test, and still no one has been able to help us. We've been told she has an atypical form of dementia, which is progressive and untreatable. We just can't give up on her yet. Can you please try to help us?"

I hated to discourage them, and I certainly didn't want to give up on her, but I had reviewed all of Martha's records. Her testing had been very extensive, and yes, inconclusive. What could I possibly have to offer her? I began with another detailed history.

Martha's daughter, a middle-aged, athletic-looking woman, spoke next.

"We've told every doctor all of this before," she said. "Can't you just look at her records?"

"Yes," I responded, "but it would be helpful to me if you would tell me yourselves, and please don't spare any detail."

Martha's husband and three grown children all contributed to describing their mother before she began to show signs of dementia. Martha was extremely functional prior to becoming sick two years earlier. The mother of six children, she managed the busy household, and in her spare time worked as the bookkeeper for her husband's store. She was engaging, humorous, intelligent, and meticulous. She never made a mistake with the books, and the household ran flawlessly. There was no family history of neurodegenerative disease. It all seemed to happen quite suddenly, although they all admittedly went about their busy lives and didn't even notice the subtle signs earlier on.

Then, one day, Martha failed to get up early as usual. When someone finally roused her, she wandered aimlessly into the kitchen and wasn't able to figure out how to open the refrigerator, or how to start the microwave. Then she picked up something to read and they realized it was upside-down.

She deteriorated from there until she was unable to even care for herself. When she talked, she appeared confused, and over time she became less and less verbal. She now spent most of her days sitting in a chair, staring out the window, often failing to even dress for the day. She had multiple brain MRIs, EEGs to evaluate for the possibility of seizures, vials and vials of blood work, and a spinal tap, to no avail.

As I stood looking at her, Martha appeared to be older than her stated age, with deep facial lines and gray hair tied neatly in a bun, dressed in a housedress that was too large for her. She just sat there, despondent, looking blankly at me when I addressed her, completely mute. She didn't follow complex commands, but inconsistently she would follow a simple command such as "Close your eyes." Her neuro exam was otherwise nonspecific with normal strength, sensation, and reflexes. Her gait and coordination were also normal. If I hadn't seen the normal MRIs, I would have suspected a frontal-temporal form of dementia that can lead to mutism, but there was no predictable atrophy in the frontal or temporal lobes.

Finally, I asked about a history of depression in Martha, or anyone else in the family, and they all denied this. Her son, a middle-aged, well-dressed businessman, addressed this issue best.

"Mom was always the upbeat one," he said, "the one that kept us all together and cheered us on when we had problems."

I just knew I was missing something important here. But what?

"Tell me again how this started," I said. "Was there anything else going on at the time that might have precipitated this?"

They all looked at each other quizzically.

"Not really," the son said. "Dad had just retired. It's a good thing, too, because he's around now to help with Tommy."

My ears perked up. "Who is Tommy, and why did your dad have to help him?"

"Tommy's our oldest brother. He was in a devastating motorcycle accident when he was just twenty that left him quadriplegic from the neck down. Mom refused to put him in a care facility and had been his sole caretaker—bathing him, feeding him, and being his source of emotional support ever since. Tommy is still at home with Dad. So, you can see how difficult it is for him, as he now has to take care of both of them, Mom and Tommy."

Could it be? Could this woman have been so needed by her son and everyone else that she couldn't stop to take the time to grieve what she had lost? What her oldest son, whom she loved so dearly, had lost? The enormity of the task of being his sole caretaker, not just for his physical needs but for his emotional needs as well, was beyond my comprehension as a mother of a son myself. Not only that, but she kept everyone else in the household going. That is, until the day her husband finally retired, and she knew she could safely turn that task over to someone else she trusted.

Pseudo-dementia is a shadow that mimics true dementia, clouding cognition and memory while concealing a different pain beneath. It's like a heavy cloud of depression, casting a cold, dull light over the world, turning its vibrant colors into muted grays. The heart feels this weight, but the cause remains hidden, masked by the signs of cognitive decline.

Yet, unlike true dementia, the core of the self is still intact, buried beneath layers of sadness and exhaustion. The mind feels wrapped in darkness, waiting for the clarity of understanding and treatment to break through. There is hope, for pseudo-dementia is not a permanent loss but a temporary—and difficult—detour.

Martha had instinctively postponed her grief-induced major depression, the cause of a pseudo-dementia, for all those years because she felt she had to. Now recognized, I referred her to psychiatry, and she underwent aggressive inpatient treatment with therapy and medication. Incredibly, she was back to her former self within six months, the treasured heart of the family.

Grief is a process that cannot be avoided. In the world of grief, time moves differently. Days stretch into eternity, and moments of joy are fleeting and rare. The past is a vivid tapestry of memories, while the future appears as a distant, unreachable land. The present is a slow march through a fog of pain, where every breath is heavy with the effort of simply continuing.

Yet, within the depths of grief, there is also a profound connection to the very meaning of love. It is the price paid for the gift of loving deeply. Though it may never fully disappear, grief evolves, occupying a small corner of the heart, a reminder of the way things were and the resilience of the human spirit.

This story teaches us that no one can escape the eventual showdown with Grief, no matter how long they manage to postpone the meeting. He demands your full attention before you can heal, and he will wait as long as necessary for you to look him straight in the eye.

PART FOUR

Private Practice

The Art of Diagnosis

CHAPTER 13

The Soccer Mom's Double Life

On the surface, she was the perfect soccer mom—carpooling kids to practice, cheering on the sidelines, and hosting weekend barbecues with a smile that never faltered. Living in a picturesque suburban neighborhood with her husband and three beautiful children, her life appeared nothing short of idyllic. But behind the cheerful facade and neatly organized routines, she harbored a secret. Every day, she leaned on something darker to get through—the steady, numbing escape of nitrous oxide. While juggling school pickups and family dinners, she was quietly battling an addiction that no one suspected, not even those closest to her. This is the story of the soccer mom's double life.

I envied Eve, a healthy thirty-eight-year-old mom of three seemingly perfect children who lived the perfect life in a wealthy suburb, driving a white Range Rover to her kids' soccer games. She was able to stay at home and enjoy those precious years of raising her children. Yes, I had my career, which I worked hard for and wouldn't have given up for the world, but there were days when I wished I could have attended all my son's practices and games and not be so tired much of the time when I was at home. No one ever said it would be easy. But for Eve . . . well, she had it all.

Or did she?

Eve first presented to me for a neurological consultation with complaints of fatigue and tingling sensations in her feet and hands, mostly at night. Her young age flagged potential concern for multiple sclerosis (MS), a disorder of the central nervous system that most often strikes the young. Her history failed to reveal anything unusual. There was nothing significant in her family or her prior history and she had previously been healthy.

She neither smoked nor drank alcohol, she exercised regularly, ate a healthy diet, and she slept well. She denied any symptoms of depression, anxiety, or stress. Routine blood work all came back negative (including a complete chemistry panel, CBC, B12 levels, thyroid studies, and autoimmune panel), and her neurological examination was completely normal. I advised Eve that if her symptoms persisted for more than four weeks, I could perform electrodiagnostic testing (a shock test of the peripheral nerves). Or, if they became more persistent, or she developed actual sensory loss or weakness, I would proceed with brain and spine MRIs.

Feeling reassured, she left, and I heard no more from her until about three months later, when she scheduled a follow-up.

"My symptoms are worsening," she said. "I'm getting the tingling in my hands and feet throughout the day now, and it's starting to keep me up at night. I'm getting worried."

She claimed not to have experienced any weakness, loss of balance, change of diet, or change in bowel or bladder function. However, her examination now revealed some diminished proprioception (awareness of position or movement) in her feet and ankles, which can be a sign of peripheral nerve damage. I repeated all her blood work, and added additional testing for anything known to cause neuropathy, and performed electrodiagnostic testing, which confirmed my exam findings of peripheral nerve damage. Despite having localized the problem to the peripheral nervous system, I scheduled brain and spine MRIs to evaluate the central nervous system. The results of those studies showed abnormal signaling, suggesting scar tissue developing in the posterior area of the cervical spinal cord.

Eve was now showing all the signs of subacute combined degeneration of the spinal cord typical of B12 deficiency, despite initial normal B12 levels in her blood work. The most common cause of B12 deficiency is from malabsorption of B12. A relatively uncommon cause is nitrous oxide toxicity, which decouples B12 in the bloodstream, inactivating it. I ordered more sensitive blood work to try to sort the cause out, and what came back was not what I was expecting.

Eve's clinical picture was now suspicious for nitrous oxide toxicity.

I had seen this condition before, typically in patients with recent clinical exposure or substance addiction. Eve was the last person I would ever imagine to be addicted to nitrous oxide, commonly known as "laughing gas," often given in the dental office with a euphoric result aimed at minimizing the patient's anxiety. If given to a patient with borderline B12 levels, it has been known to precipitously drop the levels enough to induce a severe deficiency.

But Eve denied any recent clinical exposure, leading me to conclude that only sustained significant use from addiction could cause such devastation.

It was time for me to confront Eve.

"Eve," I began, "your condition appears to be a direct result of nitrous oxide intoxication causing B12 deficiency. I'm going to treat this with B12 shots. It's imperative that you cease all use of nitrous oxide, or this will return and worsen. It can cause permanent damage to the spinal cord, brain, and peripheral nerves, resulting in difficulty ambulating, severe pain, sensory loss, and even dementia. If you stop now, I think things can still resolve."

Eve sounded surprised when I confronted her. "I don't know what you're talking about, Doctor. I have never used nitrous oxide."

Dealing with addiction is one of the most challenging tasks for a physician. With most diseases, patients are motivated to get better, more likely to follow your advice and cooperate with treatment. But addiction is different by its very nature—those affected continue compulsive behaviors even when they know the harm it's causing physically, mentally, and socially. Addiction alters the brain, leaving individuals unable to control their actions. They lie, steal, and manipulate to feed their habit. Eve was lying to me, and without her permission, I was powerless to reach out to her family. Typically, things have to get much worse before an addict seeks help for the root problem. All I could do was warn her about how devastating a B12 deficiency could ultimately be on her nervous system.

Despite her denials, Eve was treated with B12 shots and her symptoms cleared up nicely. Finally admitting to me that she may have used nitrous oxide recreationally on occasion, Eve downplayed the severity of this and was steadfast that I not discuss her condition with her husband, or anyone else, insisting she could manage this on her own. Unfortunately, the addict living this kind of double life is perhaps at the greatest risk of relapse because no one can do this alone.

I was disappointed when Eve returned one year later, this time as a hospital admission with permanent myelopathy (scarring and atrophy) of her spinal cord, unrelenting burning pain in her extremities, and cognitive decline. She finally admitted to the use of seventy-five to one hundred whippet cans of nitrous oxide daily just to get through her days.

Subacute combined degeneration of the spinal cord from nitrous oxide abuse is a devastating condition that unfolds quietly within the body. Like a hidden threat, it slowly erodes the spinal cord, leading to gradual destruction. Nitrous oxide, initially a source of fleeting joy, turns harmful as it strips the body of B12, a vital nutrient for the nervous system. Without this essential component, the peripheral nerves, spinal cord, and brain begin to deteriorate, and as the abuse continues, irreversible damage sets in.

The first signs are subtle—fatigue, weakness, and tingling. These early symptoms are faint warnings of the greater damage to come. Over time, the condition worsens, the spinal cord unraveling as it spreads its harm throughout the nervous system. The limbs, once steady and agile, begin to falter, and the simple act of walking becomes an unsteady, uncertain journey. As the degeneration deepens, the victim is plunged into sensory deprivation. The senses grow dull, and the hands struggle with diminishing perception. The world, once vibrant with colors and textures, fades, leaving only traces of sensation behind.

Without intervention, the brain, too, is affected. Speech becomes hesitant and broken, a shadow of its former fluency. The mind, once sharp and curious, sinks into confusion and cognitive decline. In the end, the spinal cord is left frayed and damaged, a stark reminder of the devastating effects of nitrous oxide abuse.

The once harmonious rhythm of life ends in a discordant and tragic note.

Eve's is a tragic story of a woman who outwardly seemed to be a happy, affluent wife and mother living the perfect life in the suburbs, while her concealed nitrous oxide addiction was causing the full spectrum of subacute combined degeneration of the spinal cord. She remained steeped in denial, despite counseling and treatment. Her story serves as a haunting reminder of the devastating consequences of substance abuse that takes over the brain's pathways with a craving so irresistible that one would risk absolutely everything, despite a seemingly perfect life.

The addict, such as Eve, forced to live this double life is often the most difficult to reach because of the solitary nature of their existence within the realm of addiction where there are no checks and balances on their behavior, and no desire to give up their daily escape into euphoria.

Eve's addiction, now out in the open to friends and family, will be watched more carefully. Sadly, she experienced the worst-case scenario from her unchecked B12 deficiency—dementia—so she may forget why she used nitrous oxide to begin with.

CHAPTER 14

When Innocence Turns Deadly

The day started like any other, with an early rise to begin my hospital rounds before 7:00 a.m. As a seasoned neurologist, this was one of those intense weeks where I was on hospital call—seven straight days of consults and admissions instead of my usual office routine. It was a warm, clear summer morning, though I barely noticed. My pager was already demanding my attention, pulling me in different directions. By now, the calls had become familiar—strokes, seizures, changes in mental status, and the occasional Parkinson's patient needing medication adjustments before surgery. I had learned to move through these cases with practiced efficiency, constantly triaging and recalibrating my day based on urgency. But through it all, one thought kept me grounded: the hope of making it home to my family at a decent hour.

Today, however, was somehow different. As I made my way around the hospital, I heard whispers of a particularly tragic case that had come in earlier that day involving a young boy who had been admitted to the intensive care unit. As a mother of a young boy myself, I hoped they would have no need for my services for this case. When a neurologist is summoned to the ICU for a consultation, it can be a sign that all the options have been exhausted. Our role is to evaluate the most severe brain conditions in cases of stroke, trauma, coma, or drug overdose. We are often the last to assess, prognosticate, and—most challenging of all—talk with the family.

The ICU consult was called in to me late that evening. What had led up to this patient's accident would remind me of the pleadings of my own eleven-year-old son when he really wanted something badly, and Matt was just about my son's age. We all can recall moments of pure joy we experienced in the innocence of youth, such as the excitement of Christmas morning, exhilarating amusement rides, dazzling fireworks, or welcoming a new puppy. For Matt, I would later learn from his family, that moment of sheer happiness occurred the day his parents finally acquiesced and presented him with the BB gun he had been longing for on his birthday. He had pleaded for months, but his parents were reluctant, due to concerns for his safety. "Please," he implored. "I'll only use it in our backyard, which is surrounded by the woods, and I'll only aim it at the targets near the barn." So, when he unwrapped the package on his twelfth birthday, he was overjoyed to find a sleek red BB gun with a stylish silver handle just like the one he'd admired in the store. He could hardly wait to show it off to his best friend, Josh.

And so this tragic story began.

I had a terrible sense of foreboding that evening as I made my way to the ICU. To the casual observer, the ICU appears an enigmatic space, often located at the hospital's core, secured by heavy glass doors that demand a magnetic card for access. Adults clad in lengthy white coats enter and exit the ICU around the clock, discreetly slipping inside as they swipe their magic cards. These doctors can be seen clustering together, murmuring in hushed tones as they navigate the circular space bordered by individual rooms concealed behind expansive sliding glass doors. They can be observed flitting in and out of the rooms, jotting notes on computer terminals. The air is permeated by a sterile, antiseptic scent, while a continuous hum emanates from the myriad of machines that, despite operating independently, produce a collective sound akin to an orchestra rehearsing with intermittent pauses and restarts. The occasional alarm sends the white-coated figures sprinting, as a group, toward the source of the disturbance, like white blood cells chasing a virus.

I sensed an unusually tragic atmosphere as I approached and noticed fifteen to twenty distraught family members and friends gathered outside the ICU, their faces etched with terror and streaked with tears. They watched me suspiciously as I passed quietly by, slipping through the glass doors with my magic card. They must have suspected I was the neurology consultant they had been waiting for, the final doctor to examine their loved one. I was the one, they were told, who might provide them with more information about the prognosis they were desperate for.

Only this evening, I already feared the worst, and the situation lived up to all my expectations. Just behind the sliding glass door, Matt's seemingly lifeless body rested on the hospital bed, connected to monitors and numerous intravenous lines delivering glucose in saline and blood pressure medication to treat shock, while the mechanical ventilator steadily pumped oxygen into his lungs. *Whoosh . . . click . . . whoosh . . . click.* The room was uncomfortably warm, but the humming sound of the machines offered a reassuring indication he was still clinging to life. His face and neck were terribly swollen from neck trauma, making it difficult for me to manually open his eyes. As I stood over his flawless young body, I was reminded of how fortunate I was that my own son was safely asleep in his bed at home.

I began my examination. Matt was unresponsive to my voice. He did not wince, pull away, or exhibit any increase in heart rate when I applied painful pressure to his fingernail beds trying to stimulate some response of awareness. His pupils were unreactive to light. He did not over-breathe the ventilator, indicating he had no respiratory drive. He was not sedated, and his temperature was normal. All preliminary findings were consistent with brain death. The nature of the head trauma and his young age dictated that we must allow a period of time to ensue and repeat the brain death exam before officially declaring him so.

What transpired next was too painful to describe. I spoke with Matt's mother and father, attempting to explain as empathetically as possible that the prognosis for any meaningful neurological recovery was dismal, that the doctors had exhausted all efforts, that arriving at the hospital sooner would not have made any difference, and that their son could not feel pain. In a situation like this, at the end of the conversation, I must ask the most challenging question: "Do you want to consider organ donation?" Their eyes, however, seemed to glass over after I mentioned the dismal prognosis, as if they heard nothing after that.

There is perhaps nothing more tragic than the senseless loss of a young life from an accident that could have been avoided. In such cases, a specific length of time is required, in the hope of some recovery, before withdrawing care. In the following days, quiet voices could be heard at Matt's bedside as his father talked to his unresponsive son for hours about all the things they would do together when he recovered, and all the wisdom he still wanted to impart to him about life.

The ICU staff relayed that during the late night and early morning hours when no one else was around, loud sobs could be heard at Matt's bedside as Josh wept, apologizing for that fateful evening just before sunset. The two friends had been walking back to the house from the barn, where they had been playing with the BB gun. Josh had playfully snatched the gun from Matt, aiming at his friend's neck without any intention of firing it. He was unaware that there were still pellets in the gun. He couldn't have known that it would cause a fatal dissection of the carotid artery, the major blood vessel supplying blood to the brain, leading to a critical loss of oxygen and, ultimately, brain death.

The muffled sounds of his loved ones' words and sobs were heard in the ICU, day after day, night after night, until the machines were finally silenced as the inevitability of death was realized.

As I tiptoed into my own son's bedroom that night, I kissed his sleeping head, praying he would always be spared from harm, as senseless accidents can happen to anyone.

CHAPTER 15

The Complicity of Parents

In the world of youth sports, parents often find themselves torn between nurturing their child's athletic dreams and protecting their well-being. While coaches and trainers play key roles in managing injuries, parents are the first line of defense when it comes to their child's health. Yet, too often, the pressure to succeed—whether from external expectations or personal ambitions—leads parents to overlook or downplay the severity of sports injuries. This complicity, whether intentional or not, can have lasting consequences, as children are encouraged to push through pain, risking long-term damage in the pursuit of short-term glory. Understanding the crucial role parents play in managing sports injuries is essential for safeguarding young athletes' futures.

Bobby, a bright twelve-year-old aspiring baseball player, was a member of a select travel team, arriving at the district fields for a regional baseball tournament to determine which team would advance to the Pennsylvania state championships.

It was a warm day in July and the sun, high overhead, reflected off the boys' helmets as the team lined up for batting practice before the start of the first game of the tournament. Having been first runner-up to the state champions the previous year, their hopes were high. They were proudly wearing their new black-and-white pinstriped uniforms, paid for by their parents, some of whom had difficulty affording the special gear but voted to do it anyway as their boys were special.

There was always a hint of competitiveness in the air amongst the parents. This team was composed of the most talented Little League players drawn from three townships, who had tried out and made this team. Because this was a competitive league, playing time on this select team was based more on ability and favor rather than equality, despite their young ages. Their parents wanted them to win. We all wanted our sons recognized. Bobby's dad had been a professional minor league player, and he just knew his son had the talent to go all the way to the majors someday. Attending as many games as I could, I had gotten to know Bobby's parents well, as my own son, Rory, was also a member of the team. Our boys were friends, both talented at their positions—Bobby played first base, and Rory was a pitcher—always butting heads and joking around in the dugout, as twelve-year-old boys are inclined to do.

Sitting with the other friends and family of our team in the bleachers, I relished this time watching Rory's sports. It was a welcome escape from the antiseptic halls of the hospital, where I spent so many hours attending and consulting on patients with neurological trauma and degeneration. There was none of that here. This was going to be a glorious day, as we sat watching our boys have the time of their lives!

The start of every year's baseball season was heralded by the croaking of the frogs in the pond in front of our home, marking the start of their own mating season. There was a particular enchantment to the baseball field, with fresh white chalk on the baselines and batter's box, and the clacking of the players' cleats on concrete. The smell of the concession-stand popcorn and hot dogs mixed with the sound of the boys' excitement and laughter as they got in line for batting practice, just before the start of the season's first game. I was taking this all in when I suddenly noticed a stray bat fly back and hit Bobby—*whack!*—right in the side of his red helmet, knocking him to the ground unconscious. He had been looking up at the bleachers when it hit him. His coach ran to his side. He was only out for a minute, but I saw him get up and take a few steps with an unsteady, staggering gait. As a neurologist, I knew this was not a good sign. There was much confusion as his parents ran down to the side field where the boys had been practicing.

I never know what to do in these situations. Should I intervene and offer my help as a neurologist? Surely, they would follow protocol and remove Bobby from play until he was cleared by a health professional. Or so I thought. But this was the year 2000. Traumatic encephalopathy from repetitive brain trauma did not begin to receive extensive attention until 2005. It wasn't until 2007 that the NFL would launch their "Heads Up" program, dedicated to teaching young athletes, coaches, and parents about concussion safety and prevention.

The Centers for Disease Control and Prevention didn't publish their first-ever evidence-based guidelines for youth sports–related concussions until 2012.

As I mulled over what I should do, I witnessed that the coach, after conferring with Bobby's parents, just put him back into the game. This was wrong on so many levels! Perhaps this coach was just ignorant of the damage a single concussion could do to the young, developing brain, which is particularly vulnerable to even minor trauma. OK, I told myself, if this were my son and there was a trained neurologist on-site, I would very much appreciate their help and advice. So down the bleachers I went and approached Bobby's parents. I explained to them as cautiously as I could what I had seen, and that Bobby almost certainly had a concussion.

"I'm not sure what the coach's thinking is here," I said, "but as a board-certified neurologist, I feel I should advise you about the proper management of a sports-related concussion. It's important to manage these kinds of injuries carefully, especially in youngsters. The recommendations are to remove them from play for a period of time to give their brain time to heal, and to return to play gradually, and only when they are free of any symptoms."

It was at this point that I took it to the next level.

"I don't wish to alarm you," I continued, "but there is something called second impact syndrome, which is unique to the developing brain in which the injured can develop severe brain swelling if they suffer a second impact before they have fully recovered. This can be very dangerous and, in some cases, even fatal."

To my amazement, both parents just glared at me. Bobby's mother was the first to speak.

"You're just saying that," she said, "because you think if Bobby sits out, your son will get more playing time. We're keeping him in the game."

I was momentarily dumbfounded, partly by the lack of logic in the accusation: our sons didn't even play the same position. And partly by her stupidity and seeming lack of concern for her son's well-being. Stunned in the moment, and realizing I wasn't going to win this battle, all I could think to say was, "I have done what I can on your son's behalf. It is your choice."

And then there was Tyler, a talented high school junior, and a key player for Central High in an away high school football game that would determine the regional football champions of that year in Western Washington. It was a clear, crisp Friday evening in the early fall as I climbed into the bleachers of our home team, the Raiders. They were already full of friends and family, all wearing red in support of our team. The Raiders were led by my own son, Rory, captain and quarterback for the team. The giant floodlights were shining brightly on the field, where the players were lining up for the first play. Unless I was on duty at the hospital, I never missed one of Rory's games. It was now 2005 and there was a little more public awareness about sports concussions, so the coaches should be more mindful of such injuries. Or so I thought.

The boys were playing their hearts out. At the beginning of the third quarter, the score was tied. Tyler, rushing in to receive the ball, was tackled from the rear. I watched him go down, his helmet colliding forcefully with the ground. He didn't appear to lose consciousness, though, and he managed to get right up and run to the sidelines with good coordination. Perhaps he was mildly concussed. You only need sufficient mechanical force to stun the brain and feel dazed to suffer a concussion; loss of consciousness is not an absolute requirement. On the other side of the field, I saw his coach talking with him. Then, to my surprise, Tyler ran right back in to complete the play. Five minutes later, he collapsed unconscious on the field, without a hit or any other direct cause. I feared the worst as the ambulance raced him off the field with alarms blaring and red lights flashing.

Tyler was in the OR within the hour for emergency management of severe brain swelling because of *second impact syndrome*. This is a life-threatening condition requiring a decompressive craniectomy, in which part of the skull is removed to try to relieve the pressure of the swelling brain to minimize brain damage. I later learned that Tyler had suffered a concussion just one week earlier in practice, something his coach was aware of. His coach was also aware that Tyler was symptomatic, suffering headaches all week leading up to this game. And yet he still practiced him every day. The coach needed him. Tyler was one of his most valuable players, and the championship was at stake.

Sadly, the coach was also his father.

As a neurologist, I was one of the health professionals tasked with clearing players with head injuries to return to play. This did not make me very popular with the players, who often exhibited warrior mentality. Sadly, all too often, I found that their parents could be even more insistent on their children playing—perhaps living vicariously through them. When an opportunity to protect this precious organ—the brain—was bestowed on me, I did whatever I could to make certain the child in question had fully recovered before allowing them to return to play, even if that meant they must sit out the entire season.

As a mother of an athlete myself, I also realized that we can't keep our children in glass cages, particularly the uniquely talented ones whose hearts are set on sports. Yet it should be even more concerning that some coaches and parents may be putting the games ahead of the safety and well-being of our kids. We must remember that these are just games and prioritize the future of our children above all else. Sadly, this is often realized too late.

As the years have gone by, I learned that Bobby made it to the majors in baseball. Good for him! Sadly, Tyler did not fare as well. Although he was one of the lucky ones to survive second impact syndrome—thanks to a timely neurosurgical procedure—I heard that he never played sports again. In fact, he never returned to school, having suffered irreparable brain damage.

CHAPTER 16

Difficult Conversations

How do you tell someone they are going to die? To be the bearer of that news is to hold in your hands the power to strip away the fragile illusion of immortality. You are not only telling them that their life will end, but that there is nothing to stop it, and that the road ahead will be filled with suffering. In that moment, the weight of your words is almost tangible—each syllable heavy with the gravity of finality. Time seems to pause as the reality unfolds, and the air thickens with the shared knowledge of what is to come.

Neurology is not what anyone would call a "happy specialty." We are faced daily with illnesses that threaten life, dignity, and hope. Yet, over the years, I've learned an invaluable lesson: though we may not have all the answers or a cure, there is immense power in offering compassion as we walk alongside our patients through their final days. These conversations are delicate, requiring us to not only understand the vulnerability of the person before us but also anticipate the flood of emotions that follow such devastating news. Some patients may expect it; others are blindsided. I've learned to tread gently, listening to their questions and watching their body language—the flicker in their eyes often tells me how much they are ready to hear. In those moments, more than ever, we must offer understanding and presence, because sometimes that is all we have to give.

Amyotrophic lateral sclerosis (ALS), otherwise known as motor neuron disease, or Lou Gehrig's disease, is a dreaded progressive neurological disorder that gradually robs its victim of motor strength in the limbs, speech, and swallowing muscles, and eventually the ability to breathe, as it inches its way into the respiratory muscles, causing certain death. An insidious thief, it takes up residence in the spinal cord, killing off motor neurons one by one, robbing its intended muscles of innervation. As the motor neurons die, the muscles they control weaken and eventually waste away, leading to paralysis and ultimately death. In the beginning, all this is unknown to the victim, until they begin to drop things or find themselves no longer able to climb the stairs. The cause of this disease is believed to be both genetic and environmental. There is no cure for ALS, but there are medications that may help temporarily by mildly slowing down its progression.

The spinal cord motor neurons—the ultimate target of ALS—are essential players in the orchestra comprised of all the individual motor fibers that make up a given muscle. When we want a muscle to contract, our brain instructs thousands of these individual motor neurons in the spinal cord to fire in an orchestrated fashion, resulting in an exquisitely coordinated symphony of motor units directing the individual motor fibers in a muscle to contract simultaneously.

We are born with a finite number of these spinal motor neurons as part of our central nervous system. They drop out slowly, one by one, as we age and are not capable of regeneration. When a muscle fiber loses its innervation, it shrivels up and dies, resulting in a gradual loss of muscle bulk. This is why a person seems to weaken and finds that he or she can no longer build muscle with exercise as they get older. It's the reason an eighty-year-old man can no longer build strength with training as he could when he was young.

ALS hastens this process with rapid death of the motor neurons by a programmed destruction, resulting in weakness and atrophy of random muscles. From the time of diagnosis, death typically occurs in less than five years. ALS strikes four out of every one hundred thousand Americans. It can strike at any age, but the risk increases with age and is greatest after the age of fifty. ALS is 20 percent more common in men than in women. However, with increasing age, the incidence of ALS is more equal between men and women. About 90 percent of ALS cases occur without any known family history or genetic cause. The remaining cases of ALS occur as a result of a mutated gene, which can then be inherited.

The clinical picture of this disease is unmistakable. Because it affects these unique cells in the spinal cord, it presents with unique clinical signs seen only with both upper and lower motor neuron lesions characterized by muscle fasciculations (tiny, random muscle twitches), muscular weakness, and atrophy with pathologically brisk reflexes. No other disease of the muscles will do this.

Despite the near clinical certainty at presentation, it is standard to coordinate an extensive array of tests to eliminate all conditions that could possibly mimic this certain death sentence before we sit down to have the dreaded conversation with the patient.

Patients, in turn, each have their own way of dealing with the possibility of lethality. Some, sensing something is terribly wrong, finally come in when they can no longer find an excuse for the atrophy of a limb, the inability to button their shirt, or just trouble talking, as if their tongue is thick. In some cases, fatigue alone is the presenting symptom.

Henry and his wife, Anne, were such a couple. They delayed the neurology consultation, hoping the symptoms would spontaneously resolve. She was a nurse and he an engineer, both retired. Henry, who had just turned seventy, was a short, balding man with a big smile and jovial countenance who told his story with a definite slur to his speech. Anne, an older woman with short blonde hair, sat by Henry's side in a protective manner, frequently contributing to his recent history in an animated fashion, explaining how they had researched his speech difficulty and come to some conclusions on their own. They both relayed how Henry's speech was becoming more and more dysarthric (thick), especially during the evening hours, and he was not one to touch alcohol. They put their heads together and were certain he had had a small stroke. They were optimistic that all he needed to improve was speech therapy.

Hoping for this myself, but concerned about the progressive nature of his speech difficulty, which was not typical for a stroke, I gave him a careful examination with particular emphasis on inspection of his limb muscles. And there it was: tiny painless twitches—what we call fasciculations—in various muscles including his tongue. This was a clear neurological sign that something malicious had taken residence in his spinal cord, as he was already showing signs of early denervation. If this was from degenerative disc disease, he would have had pain and depressed deep tendon reflexes, but he was asymptomatic with regard to pain at this point, with unusually brisk reflexes.

Not wanting to worry them a single moment sooner than I absolutely needed to—mindful of the fact that there was no effective treatment for ALS—I assured them that I would start with a brain and spine MRI, in addition to some blood work. I realized that they must have been aware of ALS as a possibility, as they had done their research before they came to me and Anne had a medical background, but they never asked about it. This, along with the prolonged time it took them to even seek out a neurologist, said it all. They were not yet ready to hear the news. When his testing was normal, as anticipated, I proceeded with electrodiagnostic testing of the peripheral nerves, which confirmed what I had suspected.

With every possible diagnostic possibility eliminated, I sat them down to have that difficult conversation. "I'm so very sorry," I said to them. "I, too, had hoped this was a small stroke, and that I could just set you up with speech therapy and wish you well. I'm afraid, however, that something more sinister is at play here. Your studies confirm that there is a neurodegenerative process affecting the motor neurons in your spinal cord and brain stem. This is a progressive condition."

They didn't seem to be terribly surprised by what I had told them, uttering the more commonly used name for the disease, ALS, even before I did. It was as if they had already prepared themselves for this inevitability. They accepted the diagnosis with grace, joining support groups where they did what they could to try to bring comfort and peace to others so afflicted, and planning that long-awaited Hawaiian vacation with all their grandchildren, who laughed as they watched their grandparents trying to do tai chi on the beach. They both approached the end of Henry's life, their life as a couple, with acceptance and gratitude that they had the opportunity to have shared such a long and wonderful life together. Anne continued her work with the ALS support groups after Henry died in honor of his memory, giving comfort to the families of those affected.

Miriam, who had just turned fifty-five, was different. She was tall, slender, and well dressed. Her short brown hair was stylishly cropped, and she carried herself with dignity despite appearing anxious. She presented to me for consultation when she was no longer able to reach behind her back to zip up her dress and noticed occasional slurring of speech when she talked for any prolonged period. As a teacher, she was particularly aware of this, and was concerned that this was happening more and more frequently. Her symptoms, suggestive of motor neuron disease, warranted a careful inspection of all her muscles.

I was disheartened to observe random fasciculations, those telltale muscle twitches indicating denervated muscles, especially noticeable in her tongue, and characteristically brisk deep tendon reflexes, indicating the motor neurons in the spinal cord were probably the culprit of the denervation. As hard as I tried to distract her from the worst possibility by first describing the testing I was going to order, she anxiously insisted that I discuss every single diagnostic possibility in plain language, along with the reasoning for each and every test before proceeding any further. I can't say that I blamed her. She was meeting me for the first time, and I was a young female neurologist new to a well-known practice. She was only seeing me because my older, more established partners could not fit her in right away. So, sensing that her severe anxiety could only be allayed by being completely honest, I laid out all the diagnostic possibilities, including motor neuron disease, trying as hard as I could not to emphasize the latter, or alluding to it by the more common term ALS.

Frightened and furious, she stood up and stormed out of the office. On her way out, she stopped to talk to Naomi, our receptionist, whom she knew and trusted as a fellow member of her synagogue. Miriam told Naomi that, in her opinion, I didn't know what I was doing, and insisted that she schedule her with someone more experienced. Naomi, an older woman who had been with the practice for twenty years, and whose late husband had himself been a neurologist, replied in her unmistakable South African dialect, "I will schedule you with anyone you like, but I can tell you that as I've come to know Dr. T., I can attest to the fact that when she makes a diagnosis, she is always correct."

Sadly, in this instance I was correct, and I ended up referring her to an established neuromuscular specialist at an academic center, as it was critical that she have absolute confidence going forward to her untimely end.

More rarely, I have encountered younger individuals presenting with ALS. Some of these cases, unlike sporadic cases involving older individuals, were the result of a rare inherited gene. Susan, a nurse, and mother of two small children, was in her mid-thirties. She was petite with shoulder-length auburn hair and an engaging smile. She came to me because of isolated right upper thigh weakness that she had developed postpartum, after delivering her second child, which seemed to be getting worse instead of better. Initially thinking this was a complication of her spinal anesthesia during childbirth, she had tried to ignore it. Now, three months later, she continued to get weaker.

Bursting into tears, she related that there had been three cases of ALS in her family. One of these was her father, whom she cared for at the end of his life. I desperately did not want this diagnosis for her. I was probably in more denial than even she was.

Careful examination revealed no other signs of concern for ALS, so I scheduled the usual testing. With brain and spine MRIs and blood work all normal, I proceeded with electrodiagnostic testing of her peripheral nerves and limb muscles. And there it was—the unmistakable fasciculations in three separate limbs, not yet even visible to the naked eye but detected with the amplification of electromyography. This young mother of two was about to receive her death sentence from me that day. Sadly, before I could even say the words, she could see it in my eyes as I observed the fasciculations on the amplifier as I moved the tiny needle electrode in her muscles. Knowing her fate already, she collapsed, sobbing. I will never forget that day.

Neurologists who have cared for ALS patients on a daily basis have frequently commented on the fact that most of these patients characteristically have a "nice" personality. So much so that if a patient comes in who isn't friendly, accommodating, polite, and generous, neurologists have been known to say, "He/she can't have ALS . . . they aren't nice enough." The reason for this has been extensively evaluated with the thought that perhaps those who get the disease have some predisposing genetic factor that is seen with certain personalities. Although this has never been proven, ALS is still known today as "The Nice Person's Disease."

Although ALS was first described in 1869, it wasn't until 1941 that it became widely known internationally as Lou Gehrig's disease when it ended the life of the famous baseball player Lou Gehrig, who embodied this particular personality trait of niceness. As he was being honored on the field, he modestly said in his farewell speech, "I am the luckiest man on the face of the earth." He died shortly afterward at the age of thirty-seven of respiratory failure.

CHAPTER 17

The Invisible Thief

Who is this Invisible Thief that quietly infiltrates the bright, curious minds of our children? A malevolent force, it hides in the shadows of their laughter and play, cloaked in the carefree joy of youth. Disguised as the excitement of growing up, it whispers promises of thrill and escape, all while planting the seeds of destruction. As our children dance and dream, full of mischief and delight, the Thief bides its time, waiting to rob them of the potential that once seemed limitless. What begins as innocence under the bright sunlight of life can soon wither in its shadow. This Invisible Thief—the lure of illicit drugs—sneaks in unnoticed, threatening to turn those moments of joy into a future marked by loss and despair. The greatest tragedy is how easily it hides, unseen and unspoken, while the futures of our children hang in the balance.

The love a parent feels for their child is beyond words—an all-encompassing emotion that defies description. Yet nothing can prepare you for the profound heartache when the Thief quietly slips into your home, unnoticed, and begins to take what you hold most dear. At first, it's almost imperceptible—a subtle shift in your child's behavior, a mood change here, a sleepless night there. Then the signs grow more alarming: sudden struggles in school, unexplained emotional swings, and an increasing distance between you and the child you once knew so well. You try everything in your power—more quality time, a tutor, teacher meetings, counseling—desperate for answers. But nothing works. Maybe, you tell yourself, it's just a phase of adolescence. Maybe it will pass. But deep down, you sense something far more insidious is at play.

It is not until you get that dreaded phone call awakening you in the middle of the night that your world shatters and it all suddenly becomes clear: "We have your son at the station. He's been arrested for a DUI and possession."

The lifelong nightmare you are about to live is just beginning.

"Who is to blame?" you might ask. After all, even a vampire must first be invited in, right? The Thief is very clever indeed, for he knows that it is the developing brain of the young that is most vulnerable to his advances. All he has to do is get in once, and he makes his way right to the pleasure and reward center of the brain where he sets up residence for life, essentially hijacking it for his own devious purposes. Once there, he carves out permanent pathways to the memory and craving centers of the brain. This ensures that there will forever be a continuous loop of memory and craving for that reward. This can be so intense the victim will desire the substance that produced this craving so much that it will supersede the desire for food, water, sleep, shelter, even procreation. You could play Russian roulette with a gun to his head and he would still say, "Go ahead, shoot. Just give it to me first!" Even after years of sobriety, all it takes is the sight or smell of something, anything associated with the reward substance, and that will suddenly trigger the cascade of memory and craving leading to a relapse. Those brain pathway changes are permanent, deeply ingrained, and forever altered.

Addiction is a brain disease. It is not a personality flaw or a willful act of defiance. It is not shameful. It can happen to anyone. In fact, it happens to one in every four individuals, spanning every socioeconomic status, race, and culture. Addiction comes in many colors and forms and can be unrecognizable at times. In some cases, one addiction seems to be mysteriously cured, only to discover it has merely morphed into another addiction—perhaps one more socially acceptable, or one less easy to detect. We know by now that addiction cannot be cured, but it can be managed. In all my years of practicing medicine, I have never met anyone who wanted to be an addict.

The brain disease of addiction is no stranger to the practice of neurology. I have encountered it in all forms: in the young and the old, in the poor and the affluent, in the educated and the ignorant. It respects no boundaries. The most tragic is when it grabs hold of our young and most vulnerable, when the brain is still maturing, well into the twenties. The Thief knows this is his easiest target and takes full advantage. The inexcusable Sackler family—trained physicians no less—who pushed OxyContin on our society for the sole purpose of their own greed, fueled this terrible opioid crisis in our country, stealing the life from hundreds of thousands of our children, right out from under our very own guidance. Cases like these are the most heartbreaking and heart-wrenching recollections of my life in medicine.

Marc's story is a particularly tragic tale of the devastation of a family.

Late one evening, I was called in to the ICU to perform a consult on a young man who had been found unresponsive on the street and brought to the hospital. Perhaps it was his young age, the lateness of the hour, or the pouring rain, but as I drove in, I had a bad feeling. As I made my way through the quiet hospital corridors to the ICU in the center of the hospital, I expected to encounter the usual group of concerned family members gathered in the waiting area just outside the large glass doors separating the critically ill from the living, but it was all quiet that night. Not a soul was present. When I entered the patient's room, I encountered an extraordinarily handsome adolescent with chiseled features and a head full of brown, wavy hair lying on the bed. He didn't look any older than seventeen. If it were not for the monitors and the IV lines coursing through his veins, I could almost imagine him as a child asleep in his bed at home. Thinking of my own son, my heart ached for his parents, wherever they might be. Perhaps they were here earlier and were told to go home and get some rest. This, however, was not the case, as he was listed as a John Doe. So, he must have been found without identification, and his parents had not been identified.

There was no sign of trauma, but he appeared to have been exposed to the outdoors for quite some time. He was hypothermic and the critical care team was trying to warm his body with blankets while they infused warm saline into his abdominal cavity. His toxicology screening was positive for heroin, Valium, and cocaine, and he had track marks all over his arms and legs. The anesthesiologist had just arrived to place him on a mechanical ventilator to support his breathing, and the dialysis team was on their way to initiate hemodialysis in an attempt to flush the drugs out of his system. His organ systems were starting to fail one by one.

I determined that his brain stem reflexes were sluggish but intact. I ordered an EEG (electroencephalogram) to monitor for conscious brain activity or seizures. *But where were this boy's parents? I would be out of my mind if my child were missing.*

By morning, John Doe was beginning to stabilize medically. His EEG revealed he was in a comatose state, but his brain stem reflexes were still present. Finally, there seemed to be a ray of hope that he might make it. The medical staff had been working day and night to find out who this boy was and had finally located the person who found him and had made the 911 call. She was able to provide a name, and with that they were able to locate his family. As it turned out, Marc was nineteen, from an upper-middle-class family in the area. Since his remaining issues were mostly brain related, I was given the task of calling them.

The response I got that day was chilling. Both Marc's father and mother got on the line to listen as I carefully explained that their son had come through the worst of it, and although he was still in a coma, as long as his brain stem was alive, there was some hope that he would still regain consciousness.

"Our son died to us a long time ago, when he got addicted to drugs," his father replied dispassionately. "We put him through rehab many times, and each time he went right back to his drug dealers. The last time, we put him back on the street and told him never to return. We refuse to get involved in what happens to him now or what you do with him. We can't bear to go through that pain again."

I couldn't bring myself to judge his parents for their attitude. It was clear how deeply they had suffered to reach this point of resignation. After enduring this heartbreaking situation time and again, they must have felt utterly powerless to change the course of events. It must be like losing him over and over again, each time bringing the same crushing grief and sorrow that comes with loss.

Almost as if Marc's spirit were hovering above and could hear them, by nightfall his neurological examination began to deteriorate. His pupils became fixed and dilated, he lost all brain stem reflexes, and his EEG became electrically silent. I initiated the brain death protocol.

Just before withdrawing care, Marc's sister, Julie, arrived, requesting a few private moments with her brother. Though she appeared older, their shared features were unmistakable, even through her tear-streaked face. In the stillness of the ICU, with machines working rhythmically to sustain life, her quiet sobs echoed in the sterile room, eventually giving way to heart-wrenching screams. She railed against the cruel injustice of the Thief that had stolen her beloved brother, so senselessly, so young.

Unlike their parents, who had resigned themselves to the inevitability of his fate, Julie could not accept the finality. She seemed to grasp the helplessness Marc had endured as the Thief took control of his fragile mind, but she wasn't ready to let go. Even in the face of death, it became painfully clear that this family would find no peace.

Then there was Anna, an attractive forty-year-old woman and wife of a well-known news anchor for a large metropolitan area to which our private hospital catered. At any given time, there was a one in five chance that one of her admissions would fall during my week on hospital duty. Anna was sadly addicted to multiple types of opiates prescribed by different physicians for chronic headaches—a common scenario among the upper-middle-class opiate addicts who have the connections to get their drugs prescribed. Her husband's fame didn't help the situation, as people would bend over backward to grant him anything he asked for.

My partners and I dreaded Anna's hospital admissions because the social situation rendered us powerless to get her the help she really needed. She tortured us with her manipulative demands and poor compliance with orders. The patient and her family, refusing to admit she had an addiction problem, pressed for more drugs to manage her pain. We fought them every step of the way, trying to get them to stop denying the real problem and admit her to a detox unit instead of pushing more drugs on her. Anna masterfully manipulated them every bit as much as they enabled her. Insisting her pain was real, demanding more and more narcotics to deal with ever increasing pain, her life revolved around multiple doctor visits. When she failed to get what she wanted, she switched doctors or would get her husband to coerce them for her.

Over time, Anna's situation became so out of control that she managed to somehow have a PICC line (peripherally inserted central catheter for intravenous administration of medication) inserted so she could get her drugs into her system even faster. When her pain would spiral out of control, which would happen frequently, she would get admitted to our hospital through the ER for pain management, since her narcotics were at such a high level that they required very careful management, and her prescribing doctors were not members of our hospital staff.

The pain management team at our hospital refused her on their service as they considered this to be an addiction issue rather than pain management. And because her primary complaint was headache, her admissions always fell to the neurology service. We, in turn, evaluated her extensively for a structural cause of headache and found nothing, also concluding it was mostly addiction related. As a result, we were completely stonewalled by the patient and her family, insisting that we were wrongfully accusing her, and further insisting that we increase her narcotics during her admissions. On more than one occasion, we actually caught a drug supplier sneaking into her room with additional narcotics for her, as we were tightly controlling what she could have. In a word, Anna's admissions were an absolute *nightmare* for us.

On the night in question, it was after midnight when the call came through from the hospital switchboard. There was an urgent consult in the ICU and it was requested that I come in immediately. It was Anna. Driving in, I wondered what the urgency could possibly be for her to be admitted to the ICU. I was expecting that this was the overdose we had all anticipated, and, on my arrival, I would find her comatose on a ventilator. Anna had been admitted to the hospital on a regular basis in the past, whenever she was spiraling out of control, but never to the ICU.

The ICU, which houses the most critically ill patients, some on the brink of death, should be a quiet place at night, but something else was happening in the early morning hours that day. As I entered through the heavy glass doors, I first caught a glimpse of two nurses standing behind the monitors at the counter.

"Boy, are we glad to see you!" they chimed. They didn't need to say anything more. The entire area was hushed, except for Anna's room. As I entered her room, I thought I had just walked into a New Year's Eve party complete with balloons, flowing champagne, and about twenty visitors all laughing it up. In the middle of it all was Anna, holding court—all eighty pounds of her—sitting up in this very expensive ICU bed, monitors turned off, entertaining her friends and family.

It was clear at that moment exactly what was going on. I was called in for an urgent consult just to justify an expensive ICU admission for a VIP, not because my services were urgently needed. There was absolutely no reason for Anna to be taking up a critical ICU bed. I was so furious at not only being manipulated this way but having been awakened to come in, I could hardly contain myself.

Trying very hard to control my annoyance, I said what I had to say: "I'm sorry to see you back in the hospital so soon, Anna. As you know, I was called in to perform a neurology consult, so I must insist that everyone leave. This is an ICU. No more than two visitors are allowed, and they must be only immediate family members, and only during visiting hours, which ended long ago. The other patients need quiet, and the staff needs a quiet place to tend to them."

After all the parties had politely left, I performed yet another neurological examination with normal findings, and, as always, I gave Anna my talk about the dangers of the excessive amounts of opiates that she was consuming, imploring her to reconsider detox. Her addiction, sanctioned by her doctors and family, was clearly at end stage. Her once youthful face was now sunken and wasted with a vacant look in her eyes. My consult now on the chart, with recommendations that I knew were unlikely to be followed, I headed to the elevators, thinking of getting some much-needed sleep before I had to return here in just a few short hours. My reverie, however, was again interrupted abruptly when a middle-aged mustached man wearing a poorly fitted toupee jumped out in front of me.

"Just who do you think you are, talking to us like that?" he blurted. "You treated us like we were just anyone off the street. Don't you know who I am?" Her husband, the famous TV anchor, had been waiting all this time to ambush me, not to inquire about his wife but to assuage his bruised ego.

Forcing a calm voice, I instinctively replied, "I am *well* aware of exactly who you are. It is because of who you are that your wife, who has a serious addiction problem, is unable to get the kind of care that she really needs. The only kind of care that has any chance of helping her is a detox facility. More drugs are the very last thing she needs, and you are her biggest enabler."

He appeared stunned at first, then he simply turned and walked away, after blurting his last words to me: "We'll see about that."

I wish I could say that my brutal honesty that morning made a difference, but sadly it didn't. Anna continued to frequent our hospital for a short time until the inevitable overdose brought her in—the one that finally took her young life.

We are all complicit in the cycle of addiction—the patient, the doctors who give in to their demands, the enablers who fail to see addiction for what it truly is, and the justice system that criminalizes rather than rehabilitates, fueling an endless cycle of recidivism. This pattern will persist until society collectively acknowledges what addiction truly is: a brain disease. Until then, we will continue spiraling deeper into this abyss of despair.

Addiction is one of the most challenging diseases to manage because patients often resist treatment at every step, seemingly determined to sabotage their own recovery. But this resistance is part of the illness itself, a symptom of the disease. Society is quick to offer compassion to those suffering from physical illnesses—cancer, diabetes, trauma. Yet addiction carries a heavy stigma, one that obscures its reality. Until that stigma is lifted, addiction will remain in the shadows, misrepresented as a personal failing, denied by the privileged, and shamed and criminalized among the less fortunate. To bring those suffering into the light, we must first remove the stigma that keeps them in the dark.

CHAPTER 18

The Case of the Misfolded Protein

The screaming is what I remembered most. I could clearly hear her screams in the background as I struggled to listen to her husband's voice on the phone begging for my help. This was the third time he'd called that week, desperate for something—anything—that would sedate his wife so she could get some sleep, so they all could get some sleep. She had not slept in ten days and nothing I had prescribed had helped. I had tried absolutely everything! She was completely resistant to sedation and remained awake, in such agony that all she could do was scream.

God help her! God help them all!

Susan had just turned fifty-five when she arrived at my office with her husband for the first time. She had waited for this consultation for nearly two months. As they walked into the examination room, I could already see that something was amiss as I observed Susan, an attractive, middle-aged woman with long dark hair draped over her shoulders and bright blue eyes. She had an unusual stare, her head tilted back, and her neck appearing stiff. And she had an unusual gait—ataxic, as if she were intoxicated. Something was very wrong, and no one had been able to help them.

Susan was unusually quiet, allowing Hal to volunteer the entire history. He described a rather sudden change in Susan's mood, ranging from depression to extreme agitation. Hal no longer felt comfortable having her interacting with customers at the toy shop they had owned and managed together for more than twenty years. They first sought help with psychiatry, but nothing they were offered in the way of diagnosis or medication had been helpful.

"You are our last hope," Hal pleaded. "The psychiatrist told us there was nothing she could do for us. We needed a neurologist."

Susan and Hal were high school sweethearts, and by Hal's description, they had led a wonderful life in the small town where they grew up. They had raised four children and were very involved in the lives of their six grandchildren. Their close-knit family, admired by everyone, was now falling apart because of Susan's erratic behavior.

Susan was adopted and knew nothing of her birth parents, but her children and grandchildren were all healthy. She denied any feelings of depression or any traumatic incidents. Her appetite had diminished, causing her to drop almost fifteen pounds in three months, which was very noticeable on her slender frame. She had also been having difficulty sleeping. This was a new experience for her, and one that numerous sleeping aids had failed to improve. She had no response to various anxiolytics or antidepressants, so these had all been abandoned. Her husband related that she had been clumsy, which he initially attributed to her poor sleep, but that had been worsening by the day now. She had taken a bad fall just this morning on her way here. She had also been complaining of dizziness, and her entire family was aware of a definite cognitive decline, to the extent that they were afraid to allow her to drive or care for her grandchildren. Just three months ago, she had been friendly and happy, with no prior history of any psychiatric problems. And now this.

Susan sat quietly with a blank stare as Hal gave the history. I noticed a certain rigidity in her posture, a decreased blink rate with a staring quality (something often seen with Parkinsonian conditions), and occasional brief twitches of her muscles, first in the left upper limb and then the shoulder and head. These were classic signs of myoclonus, which is characterized by spasmodic jerky contractions of groups of muscles. I was instantly alerted to this, as it could be a sign of muscle fatigue, a metabolic or toxic disturbance, or a neurodegenerative disease. But Susan was only fifty-five!

Just then there was a loud noise out in the hallway, and Susan exhibited an exaggerated startled response—not a good sign. I was reminded of a case I saw as a neurology resident of a rare and sometimes contagious disorder called Creutzfeldt-Jakob disease (CJD), which was commonly associated with this type of response.

But CJD was so rare, this couldn't be that—or could it?

I ordered extensive testing, including a brain MRI, EEG, and blood tests for virtually everything under the sun—a complete metabolic, autoimmune, paraneoplastic (autoimmunity as a remote effect of an occult cancer), and infectious panel. Her results were concerning. Her brain MRI revealed abnormal signaling in the deep nuclei of the brain and the parietal region. Her EEG was abnormal with abnormal periodic electrical discharges, and her blood work was essentially normal. It was now time for a lumbar puncture to test for elevated 14-3-3 protein—a marker for one of the prion diseases—a rare, sometimes contagious, and rapidly fatal brain disease, such as Creutzfeldt-Jakob disease.

This is not your run-of-the-mill test. It needs to be performed in a hospital setting with strict "body fluid" precautions, due to possible contagion. It is especially important for cerebrospinal fluid. Every instrument and container used must be disposable and incinerated. And this is when I first ran into difficulty.

Every time I tried to order the test, I received a phone call from the pathology department informing me that they didn't have the capacity or personnel to perform it. I finally located a pathologist at an academic center who agreed, but it would be very expensive. The procedure room would need to be sanitized, using a strict protocol, and everyone involved in the testing would need to wear special disposable isolation clothing, like a hazmat suit that would need to be incinerated afterward. The number of people involved in testing and eventual autopsy, should the patient die, would also have to be kept to a minimum.

Susan was deteriorating quickly. She was now basically bedridden, twitching, and unable to communicate. Worst of all, she was still unable to sleep, causing severe agitation and hallucinations.

The disease was taking its final hold on her.

A prion—short for "proteinaceous infective particle"—is a pathogenic agent that is transmissible and is known to cause abnormal unfolding of cellular prion proteins found most abundantly in the brain. These proteins appear as tangled, irregularly shaped strands, twisted and knotted in a chaotic manner. Unlike the smooth and orderly appearance of normal proteins, these misfolded ones have jagged, sharp edges with small protruding spikes, with sections of the protein clumping together, forming a toxic mass. Prions are not viruses or bacteria, as they contain no DNA. They are abnormal forms of a normal protein, called the *prion protein,* that is found in cells throughout our body, particularly in the brain and other neural tissues. However, if these proteins undergo this conformational change in shape and structure, they become infectious agents. They can be transmitted by bodily fluids, and once an individual is infected, these infectious proteins dance from cell to cell, causing chaos in the intricate network of the brain. Once inside the cells, they multiply, infiltrating healthy cells relentlessly, resulting in a virtual legion of abnormal prion proteins, disrupting normal cellular functions. As the abnormal proteins accumulate, they begin to fold on themselves again and again, like a stack of pancakes, disrupting normal cellular functions. Unlike other proteins that can be degraded by enzymes, prions are resistant to degradation, and so they continue to accumulate, causing more and more destruction.

In the brain, these pathogenic proteins eventually induce a deadly metamorphosis called a *spongiform encephalopathy,* where the brain starts to look like a sponge, with holes everywhere.

Prion diseases have an unusually long incubation period, can be sporadic, iatrogenic (caused by medical accident), or genetic, due to a mutation in the prion protein. Prion disease was originally publicized when they discovered it in the form of kuru disease in New Guinea in 1957, linked to cannibalism from eating infected brains. Then there was "mad cow" disease, a bovine spongiform encephalopathy in the United Kingdom contracted by eating contaminated beef. A genetically inherited prion disease has also been described—fatal familial insomnia (FFI). FFI is characterized by insomnia, as the prions accumulate in the brain, leading to spongiform destruction and eventual neurodegeneration and death similar to what we now saw in Susan. Lastly, there is Creutzfeldt-Jakob disease (CJD) with no clear inheritance. It has mostly been described as occurring sporadically with no clear cause, although there has been some association with contaminated medical equipment.

The 14-3-3 protein in Susan's spinal fluid was markedly elevated. She had no history of travel to the UK, the origin of *mad cow disease*, or elsewhere, nor of corneal transplant, intracerebral electrode placement, or administration of growth hormone, all of which are linked to transmission of prions from contamination. The hope now was that Susan's affliction was sporadic, such as seen with CJD, and not hereditary, given how many people might be at risk of inheriting it from her in turn.

There is no cure for prion diseases, which can only be definitively diagnosed postmortem. Once a patient becomes symptomatic, death typically ensues within a year. In the days that followed, Susan deteriorated rapidly before her family's eyes. This once vivacious matriarch of this large family was diminished to a shell of her former self, tortured by her inability to sleep, her mind fragmented, thoughts scrambled, emotions unleashed, and motor functions disrupted. Once this condition becomes clinically evident, patients often expire within six to twelve months. There was nothing to do now but plan for an autopsy to give the family a better idea of their risk of inheritance. That's where the real challenge occurred, as no hospital or hospice would take her, given her diagnosis, and her loving husband wanted to care for her at home until the end.

My final encounter with Susan was an accidental one. I was called in to the hospital late one evening to perform an urgent consult. My patient was sharing a room with another woman. The two were separated by a white curtain. Just as I was leaving that night, I caught a glimpse of the woman in the other bed. It was Susan, wrapped in her husband's arms. She was rigid and completely mute, signaling the end was near. Hal didn't seem to notice me as he rocked his beloved wife lovingly in his arms, knowing she was unable to sleep.

Susan's time had finally come, and it became my task to arrange for autopsy and burial. There was so little known about these diseases—just enough with regard to contagion—and no lab or funeral home would touch her with a ten-foot pole. After many unsuccessful attempts, I started to fear that Hal was going to have to bury her himself, in a hole in his backyard, and hope for the best.

Finally, the Mayo Clinic accepted Susan's body for study, and in exchange they agreed to assist with her burial. Her autopsy was only able to confirm that she had a prion disease.

It was never identified for certain which prion disease Susan had or how she had contracted it. Although the inherited genes we know about were not identified in Susan, there was always caution that this might be a new mutation that could still be passed on to successive generations. The incubation period with infectious prion disease can be ten to twelve years, and inherited forms can take many decades to present.

In the end, the disease that took their beloved wife and mother remained a mystery. The postmortem, which they had hoped would bring answers, only deepened the family's heartache. No clear diagnosis, no definitive name—just the vague and terrifying specter of a prion disease. The only advice her children and grandchildren were left with was not to pass it on, an impossible directive that carried with it both fear and uncertainty. Her family was left to mourn not only her loss but also the uncertainty that lingered, a reminder of how fragile life can be when the answers we seek remain just out of reach. Yet, in their grief, they held on to her memory—the mother and grandmother who loved deeply, even as the unknown stole her away.

CHAPTER 19

If Only

"If only" are the words whispered when I recall this tragic story. A young, promising surgeon—someone whose hands once held the power to heal—suffered from irreparable brain damage that could have been avoided. Along the way, there were signs, subtle but clear, that something was wrong. Clues were missed, a loved one neglected to check on him, so treatment decisions were delayed. *If only* someone had listened more closely, *if only* he had paid more attention to his own symptoms and acted more swiftly, he might still be the brilliant mind in the operating room, instead of a man forever changed. This is the story of missed chances and the heartbreaking consequences that followed.

Everyone liked Andrew. He was one of the newest trauma surgeons on staff. The nurses often referred to him as the *gentle giant*. He was tall and handsome, with a friendly demeanor punctuated by a Midwestern drawl that defied the standard stereotype of a surgeon, always offering to lend a hand to anyone that needed it regardless of their job description. A few years earlier, he had fallen in love with Kimberly, the transcriptionist in his office, and was now happily married with two children. Their youngest was just three months old. Now a new father, with the added responsibilities of a new mortgage, Andrew was always eager to take on that extra shift, to reduce the financial burden on his family and the work burden of his other partners.

Working a second shift one weekend, Andrew happened to mention to his partner during a surgery that he had an earache. He'd been fighting an upper respiratory infection for a few days, but it didn't keep him from working.

His partner was sympathetic. "Let me write you a script for an antibiotic," he offered.

"No thanks," Andrew replied. "Right now, I have a raging headache and all I want to do is go home and go to sleep."

He got through the surgery—another motorcycle accident victim who would live to ride another day—and it wasn't until after midnight that he arrived home. Not wanting to awaken his wife, who would be tired after caring for their young infant all day, Andrew quietly took himself down to the basement to sleep. When his alarm went off the next morning at 6:00 a.m., he didn't wake up. Kimberly didn't even think to check on him, strange as it was that he would fail to get up for work that morning, which was so unlike him. Instead, she packed the two children up and went off to visit a friend.

Hours later, when Andrew had failed to show for morning office hours and wasn't answering his phone, the staff became concerned. One of them headed over to his house. When no one answered the door, she first tried banging on the door and finally looked through the basement window and saw him on the roll-out bed, appearing to be unconscious. Within minutes the ambulance arrived and transported him to the ER. As they wheeled him in, they were already infusing him with triple antibiotics and steroids, suspecting bacterial meningitis since he had a high fever and was unconscious. There was no time to lose. Almost simultaneously they performed a lumbar puncture that confirmed streptococcal bacterial meningitis.

Streptococcus is a formidable foe that can get into the brain by several routes. It can either spread through the bloodstream or directly into the brain through the nasal passages. It is a leading cause of bacterial meningitis in adults, and it is treatable, if caught in time. Once the intruder manages to get within the protective layers of the meninges, which line the skull and vertebral canal, breaching the body's defenses, the disease wages a cruel war on the brain. The cerebrospinal fluid, once a nurturing fluid, becomes a breeding ground for destruction. This destruction is rarely selective, causing seizures, hallucinations, cognitive loss, loss of motor coordination, loss of speech, loss of sight and hearing. Those fortunate enough to survive almost always have residual deficits.

As the neurology consultant, I rounded on Andrew daily, carefully checking for any sign of hope, any sign of recovery, but as the days wore on his chances became more and more dismal as he failed to wake up. He was in a coma for seven days. As he slowly woke up, there was very little sign of recognition in his eyes, even when looking at his wife and his parents. His pupillary responses were sluggish, and his eyes were each going in different directions, zigzagging wildly. When he was finally able to sit up in a chair, he required maximum assistance to steady himself. His limbs were so ataxic that he was unable to ambulate or even feed himself. His speech was completely incoherent. This brilliant, talented young surgeon at the very beginning of his career and his life, now reduced to a shell of a man with severe brain impairment from the streptococcal attack, was unlikely to recover.

If only Andrew had received antibiotics just a few hours earlier, he might have fully recovered. *If only* he had been less thoughtful of his wife and slept in his own bed, she might have noticed how sick he was. *If only* she had thought to check on him that morning to see if he was all right . . . perhaps. We will never know.

During the acute phase of his illness, Andrew's parents and Kimberly were all at his bedside day and night, but as time went on, I saw less and less of Kimberly. It was becoming clear that Andrew was going to require full-time care, and Kimberly was not going to be able to manage this with two small children. His parents, Ted and Nancy, sold their home in the Midwest and bought a place in Western Washington to be close to their only son, and care for him in their home if needed. The community raised a substantial amount of money for his wife to help in her need, as they thought so very much of Andrew.

After discharge, I thought about Andrew often. I imagined he was at home with his parents getting full-time loving care, and perhaps this was helping him with some modicum of recovery. I imagined Kimberly and his children were visiting him often. Then, about a year later, I came upon him while performing a consult on his roommate in the nursing home attached to the hospital. Andrew was all alone in a darkened, dreary room, staring vacantly at the wall. I said hello, but he did not appear to hear or see me. I so hoped that in the recesses of what was left of his once bright, thoughtful, and kind mind that he had no knowledge of where he was or what had happened to him. That he was at peace.

Such is the tragic story of the unexpected. It never occurred to Andrew that he could develop meningitis; he assumed it was just a harmless upper respiratory infection. In his thoughtfulness, he chose not to wake Kimberly, an exhausted new mother. Kimberly, in turn, didn't find it unusual that he hadn't risen early for his hospital rounds. These small, seemingly inconsequential decisions led to a delay in the lifesaving treatment he needed, condemning this talented young surgeon to spend the rest of his life in a dim, unwelcoming nursing home.

CHAPTER 20

Lewy Body Karma

Have you ever wondered if karma exists? Does the universe really have a way of balancing the scales of justice?

Simon did.

The human brain is made up of billions of neurons, forming a network so complex that even the most advanced computer systems would envy it. These neurons communicate through electrical signals, powered by basic elements like sodium, potassium, and calcium. These elements move in and out of the neurons through special channels, controlled by chemical messengers called neurotransmitters. This network is not only responsible for generating and storing our thoughts and memories but also manages our emotions, processes all our senses, and directs movement and speech. Like a computer, the brain needs energy to function, which it gets from glucose, nutrients, and oxygen. All this activity creates waste by-products, which the brain normally cleans up through chemical reactions. In a healthy brain, this process runs smoothly day and night.

However, because the brain is so complex, it's also vulnerable to malfunctions. Sometimes things go wrong, and proteins can build up, interfering with how neurons communicate with each other. The result is often a neurodegenerative disease. We don't fully understand why this happens—it could be due to aging, genetic mutations, or possibly exposure to toxins. The exact protein buildup causing the disease state can only be confirmed after death through an autopsy. During life, doctors rely on a patient's symptoms and compare them to known patterns of brain diseases studied over the last century. These protein buildups are linked to several different diseases, but unfortunately, we still don't have a cure for any of them.

Simon, a fifty-eight-year-old dentist, first came to see me at the end of the summer. I was struck by the presence of this tall, handsome man with graying at the temples and an engaging smile as he walked into the exam room. As he began to describe his symptoms, I was not aware of anything unusual in his facial expression, his voice, or his ability to give me a coherent history. He initially made light of his decision to seek a neurology consultation, saying that a good friend and colleague in town had suggested it, having noticed for some time there was something odd about his gait and the way he held his nondominant left upper limb flexed at the elbow when he walked. It was something Simon was unaware of himself. He was not experiencing any difficulty running his practice, he said, or doing the intricate fine motor work required of a dentist. He was only here at the urging of his friend, reiterating that he hadn't noticed anything of concern himself. I inquired as to whether his wife had noticed anything, to which he replied that he was divorced.

On examination I noted that Simon had diminished fine motor movements of his left hand, along with rigidity of that limb without evidence of tremor. His gait was mildly slow and shuffling, with poor arm swing on the left. And he tended to hold his left arm in a flexed posture close to his chest as he walked.

Simon clearly appeared Parkinsonian, although the absence of tremor was concerning for an atypical form of Parkinson's, which was not typically a good sign.

Typical Parkinson's disease usually starts with symptoms like a tremor when the muscles are at rest, stiffness in the limbs, slow movement, and a shuffling walk. Later, people may also experience balance problems. This form of Parkinson's is linked to the buildup of a protein called alpha-synuclein, which clumps together in the brain to form what are known as Lewy bodies. These were first discovered by scientist Frederic Lewy in 1912. In some cases, Parkinson's disease can lead to dementia caused by these Lewy bodies, but this usually happens many years after the initial diagnosis. Early on, many of the physical symptoms—like tremors and slow movement—can be treated with medication containing dopamine, a brain chemical that helps control movement. This is because, in Parkinson's disease, the cells that produce dopamine start to break down, causing problems with movement.

Atypical Parkinson's, however, has some differences. It may look similar at first, but in many cases, patients don't have a tremor, as in Simon's case. The damage in atypical

Parkinson's affects not just the dopamine-producing cells but also the cells that rely on dopamine. As a result, medications that work well for typical Parkinson's often don't help as much here. This form also leads to broader brain degeneration, which can cause dementia to develop earlier.

One type of atypical Parkinson's is called Lewy body dementia. Unlike typical Parkinson's, it usually starts without a tremor and quickly progresses to dementia within a year. This was the disease that comedian Robin Williams suffered from. Lewy body dementia is known for its fluctuating symptoms, including visual hallucinations, mood swings, depression, and anxiety. These hallucinations and psychiatric symptoms are often difficult to treat, as common medications for Parkinson's don't work well for them. In fact, some medications used for hallucinations can actually make the symptoms worse, leading to a frightening experience for the patient.

The challenge with neurodegenerative diseases like these is that there are no blood tests or brain scans that can definitively diagnose them while a person is alive. Doctors have to base their diagnosis and treatment plan on how the patient's symptoms progress over time. The exact type of disease can only be confirmed during an autopsy after death.

I found it concerning that Simon didn't seem to be aware of his deficits, indicating poor insight—also a poor prognostic sign. When I inquired if he had ever been told that he acts out his dreams in his sleep, he replied that this had indeed been problematic for him. It was not unusual for him to punch or kick the person sleeping with him, unintentionally. This was something known as RBD (REM sleep behavior disorder), which can also be a prognostic indication of future dementia. He had no known family history of neurological disease.

All the signs indicated that Simon's form of Parkinson's was concerning for one of the atypical Parkinson-plus syndromes with additional features unrelated to the movement disorder. The most concerning of the plus features was early dementia and hallucinations. But how could I tell Simon this without taking away his hope? He was a long way from retirement, and he told me he was engaged to be married. I started off by explaining the movement disorder part—what most people think of when they hear Parkinson's disease—advising a trial of dopaminergic medication to see if that would help. I also ordered a brain MRI and baseline neurocognitive testing. Lastly, I said, "Please bring your fiancée with you next time. I would very much like to speak with her."

Simon was clearly a very intelligent man and performed well on the neurocognitive testing initially, except for some attentional errors. His brain MRI was normal. He had no improvement in his gait or stiffness with an initial trial of dopaminergic medication, further suggesting we were dealing with something more sinister than traditional Parkinson's disease. At his follow-up visit, his fiancée, Karen, an attractive brunette, was hesitant to say much in front of Simon as they sat before me holding hands. She informed me, however, that they had been together for ten years and had finally just decided to get married. She denied noticing any cognitive changes. It was time to push the dopamine dose higher and higher to try to get some response.

Within a few months, Simon's cognition started to falter, and he was beginning to experience hallucinations, which failed to respond to antipsychotic medication. This presentation was more concerning for Lewy body dementia. The further I titrated his dopamine dose upward without a response, the surer I became that this was the case. It was now time to advise him to sell his practice and prepare his fiancée and grown children for what was to come. Simon still denied any cognitive issues, although by now there obviously were some. With his permission, I contacted the close friend and colleague who had prompted him to seek help in the beginning, and who now helped confirm Simon's further cognitive decline and volunteered to help steer the discussion with Simon privately.

As I followed Simon over time, I witnessed the gradual fading of a once brilliant mind. He had fleeting moments of lucidity before retreating into the foggy depths of confusion. His eyes betrayed the torment within, as he became lost in the disarray of fragmented thoughts. Over the ensuing months, as Simon's cognition deteriorated, he was tortured by visual hallucinations that he described as psychedelic colors, people wandering around his house like ethereal specters, animals in the shadows baring their teeth at him, and giant faces in the trees glaring in at him.

Karen had given up her job and moved in, as Simon now required full-time care. She informed me that at times he was lucid, and other times he couldn't find his way to the bathroom. And when he did, he had forgotten how to unzip his pants and use the toilet. In the early stages of the disease, she vowed to marry him and take care of him, but that was all changed now, as her compassion turned to sorrow for what was and what would never be again. She requested help with nursing home placement.

One day, Simon unexpectedly showed up at my office without an appointment. In a lucid moment, he had somehow gotten the keys to his car and driven in on his own, because he had something he just had to get off his chest.

"I never thought anything like this would ever happen to me," he said. "I'm certain this is karma for what I did to my family. Ten years ago, I had an extramarital affair with Karen, who was my wife's best friend at the time. This destroyed our marriage, and my teenage children have wanted nothing to do with me since. Even now, as I deteriorate, they refuse to see me. I postponed marrying Karen all these years, hoping to mend things with my children. I regret this will never happen now."

Sadly, as Lewy body dementia crept through his mind, blurring reality and memory, Simon couldn't shake the thought that this was his reckoning. The guilt weighed heavier with each passing day—faces from his past, people he had wronged, flashing before him in moments of unexpected clarity. Was this his punishment, this slow unraveling of his mind and body? Or was it simply chance, a cruel twist of fate?

Simon died alone in a nursing facility about a year later, never having reconciled with his family. His wife and children had moved on long before his illness took hold, but tragically, Simon never could. As his condition worsened, he was plagued by terrifying hallucinations, with fewer and fewer moments of clarity. Communication became nearly impossible as he drifted further from reality. I felt powerless to help him escape his torment—medications often made things worse, and I could no longer reach him in any meaningful way. His delusions deepened, leaving me unable to offer any real comfort.

Simon's unexpected visit to my office during one of his rare lucid moments suggested to me that, in those brief periods of awareness, he was haunted not only by his illness but also by the heavy weight of regret—perhaps even more so than by the hallucinations that gripped him.

May he finally find peace.

CHAPTER 21

Death with Dignity

The Failed Video Recording

In the prime of his life, when everything seemed to be falling into place, Jason was struck by a mysterious and relentless neurological condition. What began as a simple episode of vertigo, his symptoms soon spiraled into a debilitating illness that no doctor could diagnose, no treatment could ease. With every failed test and ineffective therapy, hope faded, replaced by a growing sense of despair.

It was a beautiful Friday afternoon in July and my thoughts were already tuned to my weekend plans to leave for the island right after dinner. It had been a long week and the day was just about to wrap up but as often happens late on a Friday afternoon, I received an emergency call. This time it was from an occupational therapist I knew from the hospital. The patient was her husband, Jason. He had been experiencing severe vertigo for three days, accompanied by nausea and vomiting, and she was getting more and more concerned. She said she could have him in my office by the end of the day, so I agreed to wait for him.

Vertigo, a common neurological complaint, plays havoc with the laws of physics, spinning its victims into a bewildering whirlwind, throwing the patient's world into a kaleidoscope of swirling motion, where up and down lose their meaning and the horizon becomes a wild illusion. Reality becomes a runaway roller coaster, while the mind struggles for stability. Nausea is an inevitable consequence, relentlessly churning the stomach contents. The world becomes colored by a sickly pallor, as waves of discomfort overwhelm the senses. Each movement, no matter how small, triggers severe disorientation and the motion sickness that accompanies it, such that one dares not move for fear of aggravating this vicious beast within them.

As I caught a glimpse of Jason, a middle-aged, athletic-appearing man, in my waiting room, I was alarmed at the severity of his symptoms. He sat as still as he could in his chair, with his eyes closed, clutching a large black trash bag that he repeatedly threw up in. That was not a good sign.

My intuitive alarms went up even further when I met with him in the exam room and we began to talk. I could tell almost immediately that he was an unusually kind soul, and somehow, I instinctively knew that he was going to have a bad prognosis. Something was terribly wrong. As an EMT himself, he could see that I was alarmed during my examination and at one point he stopped me to say, "The look in your eyes is concerning me. Please be honest. I know something is very wrong here."

He was right. His pupils were fixed and dilated, indicating brain stem dysfunction, a central cause for his vertigo and nausea. His problem was not the typical inner ear disturbance that I had expected and perhaps hoped to find. Something was terribly wrong with his brain.

I admitted Jason to the hospital immediately so that I could coordinate the necessary testing on an urgent basis and treat his symptoms aggressively to try to keep him comfortable. Brain MRI, with and without contrast; MRA (angiogram); lumbar puncture for spinal fluid analysis; blood work; CT scanning of the chest and abdomen to assess for a remote cancer causing an autoimmune attack on the brain. Every test came back entirely normal. However, in addition to the absence of pupillary reflexes I'd noticed in the initial examination, within twenty-four hours he had also developed a central form of nystagmus—erratic eye movements associated with double vision. Repeat brain MRI with and without contrast dye forty-eight and seventy-two hours later remained unremarkable.

Jason was poorly responsive to symptomatic treatment and remained deathly ill from vertigo, nausea, and vomiting, unable to even ambulate to the bathroom. After drawing every known laboratory test to assess for infectious or autoimmune encephalitis, I felt it safe to start him on a trial of intravenous steroids to try to quiet this sinister, invisible force that was invading his brain stem. This also was unsuccessful.

Finally, after exhausting everything I could locally, Jason agreed to a transfer to the university hospital where he could get more specialized care and testing. But despite this—and two additional transfers afterward, including the Mayo Clinic—his actual diagnosis remained elusive, and his condition continued to deteriorate. He underwent numerous treatments aimed at suppressing autoimmune attacks on his brain stem, which was presumably what this was, although the cause was still unclear. They tried immune suppressive agents. They injected multiple trials of intravenous immune globulin, an expensive infusion of the plasma from multiple donors, used to halt an autoimmune attack mounted by a patient's own immune system. When this failed, they gave plasmapheresis a shot—an invasive process designed to cleanse the bloodstream of antibodies. All to no avail.

Jason suffered through all of this for nearly a year, with no improvement whatsoever. After all these treatment trials, the best I could say to him was that he did not seem to have gotten any worse. But he was certainly no better.

Now back at home, he returned to me for ongoing symptomatic management and observation. His medical team had been rendered helpless in the face of this treacherous enemy, invisible to the naked eye and capable of eluding diagnostic evaluation. Having lost about seventy pounds, now with a feeding tube, he described to me how deathly sick he had gotten when he merely tried to take one bite of turkey at Thanksgiving!

Jason, now in a wheelchair, and his wife, Carol, an attractive, middle-aged woman with straight cropped blonde hair, sat in front of me looking more discouraged than I had ever seen them before. Both had always tried to maintain a positive attitude, but now they were faltering.

Jason spoke first. "I have been completely unable to walk because even the slightest movement intensifies my vertigo. Even when I try, I have no sense of balance and must hold on to the wall. Despite the tube feedings, I continue to lose weight, and my arm and leg muscles have atrophied so much,

I don't even recognize them. I've lost the job I loved; I can no longer take my dog for long walks on the beach. Will this ever end? I've consulted with many experts in the field, yet no one has even been able to tell me why this happened or if I can ever expect to get better."

Carol spoke next. "I had to take a family leave from my job at the hospital to care for Jason. We've gone through all our savings and have had to dip into our retirement. The medical bills keep piling up since our insurance denies most of the claims because Jason doesn't have a definite diagnosis. They keep telling us not to worry about it, that they will get it sorted out. That's hard to believe when we're getting calls from collection agencies."

Jason was managing the severe vertigo with the aid of Valium, a drug used mostly for recalcitrant cases of vertigo for its inhibitive effect on the brain's vestibular center. While it helped Jason manage the vertigo and nausea, he constantly had to increase the dose for it to remain effective.

This once upbeat, positive man, a health care worker himself, now broke down in tears in my office. The life he had loved was no longer worth living, and there did not seem to be even the slightest ray of hope for a diagnosis, let alone a cure.

I could see the look of utter despair in his eyes as he recounted what his life had become. Even his wife, who had supported him through all these months, had given up. She tearfully told me that they had investigated assisted suicide. At the time, this would have been legal in the neighboring state of Oregon with its "Death with Dignity" law, an embodiment of society's attempt to grapple with the complexities of terminal illness, allowing the individual to exercise agency over their own fate. This legislation was designed to protect the value and autonomy of each human being under the concept that dignity must extend beyond mere existence.

Unfortunately, Jason did not meet the strict criteria to be eligible, as he lacked a definite diagnosis. In other words, they could not determine that he was terminal, and there were no exceptions to the rule: terminal means that death can be predicted within six months.

So even assisted suicide had failed Jason.

I tried as hard as I could that day to encourage Jason and Carol to keep trying, suggesting perhaps that he resume immune suppressant treatment. Traveling had become more and more difficult for him, as they lived on one of the San Juan islands and the ferry ride to the mainland was becoming unbearable with his vertigo. They finally agreed to start an oral immune suppressant along with monitoring blood work on the island, if I could manage him remotely until he improved enough to resume the ferry rides to the mainland. I agreed.

A few months later, I received a message from Carol informing me that she had changed Jason's prescription refills to a mail order pharmacy, which would be cheaper for them. She requested that I send all new prescriptions for his immune suppressant and Valium to them for a three-month supply instead of just one month. This sounded completely reasonable, as they were weighed down with many unpaid medical expenses now totaling hundreds of thousands of dollars.

Their pharmacy request sat on my desk for days. I wasn't sure why, but I delayed signing it until finally I was leaving for a short vacation, and I was clearing things up.

Upon my return, my Monday started with a phone call from the medical examiner.

"I have quite a story to tell you if you can give me a few minutes of your time," he said.

It seemed that a patient of record who listed me as his only physician had committed suicide that weekend! From what he could tell, the patient had followed all the usual protocols to fulfill the Death with Dignity law, only there was no approved application on file.

"Do you know him?" he asked. "Were you his only treating physician? Did you supply his medication? Do you believe in assisted suicide?"

Recalling Carol's request for a three-month supply of Valium, I knew right away who the medical examiner was talking about.

"Yes," I replied. "I was his only treating physician. And yes, I supplied all his medication. Do I believe in assisted suicide? Well, that all depends."

It had been several months since I had seen Jason in the office because of the difficulty he had with travel, but he had crossed my mind more than once during my vacation, so I was planning on placing a call to his wife to inquire about him today. Now, too late, I painfully realized what they had been planning when Carol requested a full three-month supply of Valium, a fatal respiratory depressant when taken in sufficient quantities. They obviously made up the story about the new mail order pharmacy in an effort not to involve me directly.

The odd thing was, that medication request for Valium sat on my desk for the longest time. He had been on Valium to try to ease his vertigo and nausea for a year, and I had kept up with his regular monthly requests from the pharmacy. I usually took care of prescriptions right away, but there was something . . . just something that made me stop and wait this time.

The medical examiner was in a quandary as to whether to call the death a suicide or initiate an investigation. He continued, "Can you help me with this? Have you ever participated in assisted suicide under the Death with Dignity Law?"

I replied, "No and yes. I have never participated in doctor-assisted suicide under this law, but under certain circumstances I could see why this would be the humane thing to do. Having seen firsthand the harrowing journey patients like Jason were forced to travel, I see the law as an opportunity for patients, given the burden of suffering with terminal disease, to maintain some control over their lives, allowing them to end their lives with some modicum of grace and poise rather than suffering and indignity. But, as a rule, I do not in any way participate in assisted suicide in my own practice."

The medical examiner went on to relate that sometime on Saturday the patient's brother had called Best Buy, frantically complaining that the video camera they had just bought malfunctioned right in the middle of their brother's suicide, and they demanded help fixing it before it was too late! The poor person on the other end of the call freaked and called 911 and the police.

As it turned out, the local EMTs knew Jason well and they refused to respond because they also knew he was a DNR (Do Not Resuscitate), and they were not permitted to resuscitate him. They also knew how terribly Jason had been suffering and they had a clear understanding of the course of action he'd taken.

The police were not sure what to do either. At first, they refused to go. Then, by the time they finally did, at the insistence of Best Buy, it was too late. The deed was done, and they found Jason's home filled with many relatives and friends that had flown in from all over the country to say goodbye and witness his death, including his loving and devoted wife. He had specifically wanted his suicide filmed so that no one else could be blamed for his death once he was gone—this was in the days before smartphone cameras.

Jason had told everyone that the suicide had been authorized by the Death with Dignity law. Only he and Carol knew that the application had been denied.

Recalling a private conversation I once had with Jason during the many trials of his treatment, I wasn't entirely surprised he had chosen suicide by what he shared. He recounted an incident that occurred at the end of a plasmapheresis session—a procedure where blood is removed, the harmful antibodies in the plasma are filtered out, and the blood is returned to the body. During one session, Jason's blood pressure dropped so low that his heart stopped, and he had to be resuscitated. Remarkably, Jason described the experience as being transported to a place filled with radiant light, where all his suffering vanished, replaced by overwhelming feelings of love and peace. When he was brought back, he felt disappointed. Though he hesitated to call it a "near-death experience," he quietly confided, "I am no longer afraid to die."

Jason was an incredibly nice man who had a terrible disease that eluded diagnosis and treatment. As an eyewitness to the extent of his suffering and the lack of any hope for a reprieve from his torment, I understood his need to take control and end his suffering. His adoring friends and family, who were witness to his torment, understood enough to go along with his plans. I'm only sorry that the medical establishment failed him. In the end, Jason executed his plan the way he lived: with love for those he cared about, dignity, and grace.

I think Jason knew that if he had given me the opportunity to assist his suicide, I would have done all I could to keep struggling for a treatment. I was grateful to him that he didn't put me in that position. The coroner, after hearing my story, finally declared his death a suicide.

Case closed.

Whether viewed as a compassionate act of grace or an ethical travesty, the Death with Dignity law serves as a reminder that the boundaries of our relationship to life and death are not fixed, but rather they are susceptible to the tides of societal progress and the shifting orientations of our moral compass. In the face of this controversial legislation, humanity is confronted with a profound question: How do we reconcile our desire to alleviate suffering with our reverence for the sanctity of life? The answer to this question lies within the hearts and minds of each individual, and the collective wisdom we foster as we navigate the delicate dance between life and death.

CHAPTER 22

A Tight Squeeze

The human brain is an amazing organ. Billions of neurons communicate at light speed through an array of electric impulses propagated along channels with opening and closing gates, and terminating in the ultimate message via unique chemical messengers—the neurotransmitters. Our minds, which control our thoughts and actions, are less easy to define. Yet surely they exist, or we would be like AI robots and totally predictable. Our brain works in concert with our emotions, with all their complex defense mechanisms. These emotional orchestrations are different in every individual, yet they are similarly designed to protect us, allowing us to dissociate from experiences that are too terrible to handle. Some spiritual leaders link our minds to our spirits, but that's another story altogether. Suffice it to say that we, the practitioners of medicine, recognize a definite mind–body spiritual connection, irreversibly entwined, which brings me to Anthony.

Anthony was a twenty-eight-year-old man referred to me for a multiple sclerosis evaluation by a colleague who made the referral personally by phone. "I'm hoping you can see this young man right away," he said. "He's deteriorating quickly, and I've done all that I can on my end as a family physician. He's such a nice young man. I'm really worried about him. He lost his medical insurance when he lost his job, so anything you can do for him will be much appreciated. It's a very sad story."

Struck with severe muscular spasticity involving all his limbs, Anthony was so stiff that he could barely walk into my office unassisted. A good-looking young man, second-generation Hispanic, well-groomed and dressed casually in jeans and a clean T-shirt, he told me his story.

"I just woke up one day and fell to the ground when I attempted to get out of bed. My legs just wouldn't work. As the day progressed, my legs got more and more stiff, and then the stiffness started spreading into my arms. Now I can barely walk or use my hands. I had to move back in with my parents because I can't manage on my own any longer," he said.

I had to tease the rest of the history out of him, noting that he said yes to just about everything I asked him, eyeing me curiously as I asked each question. *Yes*, he had difficulty passing his urine. *Yes*, he had sensory loss in his arms and legs. *Yes*, he had pain in his muscles.

On examination, Anthony's gait was unmistakably spastic. He had the scissor-like walk that is often seen in someone with cerebral palsy—every step appearing like his body was struggling to move while being squeezed into an invisible tube. Muscle testing also confirmed severe spasticity in his muscles, something commonly seen in MS. The one thing about this nice young man that didn't make sense, however, was his flat affect and relative indifference to this devastating condition, which had stripped him of his ability to care for himself, let alone work.

"It is what it is. I can't do anything about it. I even had to give up my dog and best friend, Charlie, since I can no longer care for him," he replied with an air of indifference.

I carefully discussed the diagnostic possibilities with Anthony, which, at his young age, included MS, at the very top of the list. I described the testing that would need to be done, which included brain and spinal cord MRIs, and lumbar puncture, stressing the importance of getting the testing done as soon as possible so I could get him on the proper treatment.

With the current state of the art, there was plenty we could do to treat MS. He was perfectly agreeable to do whatever was needed. I just needed to find a way to help him hurdle the cost of the testing, as he no longer had health insurance. There are programs that help patients with MS, but that was going to be difficult to navigate until I could establish a definite diagnosis. I was happy to donate my own time and would try to get the radiology facility to do the same, as time was critical here.

At first, I was surprised when all the testing came back completely normal given the severity of his symptoms. But then I recalled how indifferent Anthony had appeared to be about all this, and it began to make sense. Suspecting there must have been some terrible trauma underlying his sudden deterioration, I brought him in for a more extensive talk.

Inquiring a little more about Anthony's life, I learned that he had worked for the city for the past eight years. His job involved going into underground pipes to repair them. He recently decided he no longer wanted to do this and, in an effort to better his life, he had quit his job and worked tirelessly to become a firefighter. Surprisingly, this young, athletic man failed his physical test three times. Anthony next applied to the police academy, but he repeatedly failed the physical part of that testing as well, for reasons that were not entirely clear to him or to me. Since this all started, everything seemed to be going wrong in his life. Even his fiancée left him. The more I probed, the more dead ends I encountered. He didn't appear to be hiding anything. He really had no idea why this was happening to him. And neither did I.

"Do you have any psychiatric history, such as depression or anxiety?" I asked.

"No . . . no . . . and no," he replied flatly.

La belle indifference is the French term meaning "beautiful ignorance" and refers to a patient's lack of concern about their symptoms despite a serious medical disorder. It is a classic finding in conversion disorder, also known as a functional medical disorder, which is a mental disorder usually caused by severe emotional stress or trauma that manifests as a physical symptom. In other words, the symptoms are very real and completely out of the patient's conscious control, but have an emotional rather than a physical cause. The mind is tricking the body. This nice young man was clearly suffering and stuck in this debilitating physical illness, and even he didn't appear to know why or what to do about it.

Now certain there was nothing physically wrong with Anthony but wanting desperately to help him, I was careful to validate his condition because, after all, his symptoms were very real to him. And he had no conscious control over them. For him to give up his dog? That was most certainly real. To try to help Anthony, I immediately contacted Dr. H., who was the best therapist I knew. She was also skilled in dealing with conversion disorder.

"Anthony has no health insurance, or any other means of payment. Can you please help him?" I implored.

Dr. H. was agreeable to help and donate her time, but just to one meeting.

After the initial meeting, I received a call from Dr. H.

"What a nice young man!" she said. "You are so right. Something traumatic must have happened to him, something he has buried so deep that he is completely unaware of it. It's going to take more than one session to help him. I don't care how long it takes. This young man deserves to be helped."

One meeting became two meetings every week for six months. I would often see Anthony struggling to ambulate with the help of his walker in the park outside my office. It was a sad sight to see this young man, once seemingly healthy and normal, now physically crippled, in obvious pain despite being treated with anti-spasticity medications.

About a year into his saga, knowing how much I loved dogs, Anthony stopped by my office one day to introduce me to Charlie, his white Lab. I was pleasantly shocked to see him walking completely normal, smiling, and proud to tell me he had been accepted into a training program to become an emergency medical technician.

"Thanks, Doc," he said. "Thanks for all your help, especially referring me for therapy. I had no recollection of what happened to me the day I decided to quit my job for the city.

It was only through intense therapy that it all came flooding back. I was sent in to fix a broken sewer pipe underground and became trapped for eight hours when there was a landslide from the storm. I was unable to move and could barely breathe. And I could feel the rats biting at my legs in the darkness. I thought I was going to die down there! Once I remembered that terrible day, as hard as it was to relive it, I started to get better. I'm still in therapy, but Dr. H. only needs to see me once every three months now!"

I think about Anthony's story every time I encounter a patient who appears to be indifferent to his suffering, realizing there is a hidden trauma in all of us. Some traumas are so severe that our brains dissociate from the memory of them to protect us. But in the end, the trauma must come out somewhere, and for some it will be a physical manifestation.

Anthony had completely blocked any conscious memory of that terrible day at work. All he knew was that he needed to get into some other line of work. So he quit, with ambitious plans first of becoming a firefighter. When he couldn't pass the physical testing, he tried to get into the police academy, but he was physically thwarted there too. His body just wouldn't allow him to move forward in life until he dealt with his trauma.

His family history also played a big role in his need to dissociate, as he came from a family that brought him up not to cry or show any emotion. The only being he felt safe opening up to was his faithful dog, Charlie, who thankfully was returned to him.

CHAPTER 23

Sorry

What can I say about a man who died as obscurely as he lived, who thought so little of himself and the life he had led that he left instructions that there was to be no notice of his death, no ceremonial good-byes, no flowers? No one even knows what became of his ashes. He died alone, just as he had chosen to live, spending his final days in the hospice ward of a nursing facility.

That had always been his fear: that he would deteriorate to the point that he could no longer move on his own, swallow, or communicate. Such is the ultimate fate of some Parkinson's patients. I had met him ten years earlier. Knowing that I was a neurologist who knew about these things, he had this discussion with me. He asked me if, should that time ever come, I would be so kind as to help him slip into the dark embrace of death with some sense of dignity. If he should ever deteriorate to this degree, he did not want to be kept alive in some state of bare subsistence, on a feeding tube spending his final days looking out a window. I understood.

And in time he came to mean a great deal to me.

While we lived a thousand miles apart, I made a point to keep in touch, visiting yearly at first. But with the insidious paralysis of Parkinson's disease taking a greater toll on his mobility, and my busy practice and young family taking up my time and attention, our communication dwindled to weekly phone calls. We talked mostly about his deteriorating condition; endless doctor visits; failed medication trials; my young son, Rory; and, of course, the weather. There always seemed to be a ray of hope in his voice as he would inquire about any new miracles on the horizon for the treatment of Parkinson's disease. Despite his deteriorating physical condition, his mind remained as sharp as ever.

Then, a few weeks went by without me hearing from him. So I picked up the phone and dialed his number. When there was no answer, I contacted the local hospital to see if by chance he was there. I was informed by the nursing staff that he had been admitted a week earlier with severe protein calorie malnutrition, inability to swallow, and aspiration pneumonia. Without his consent, a feeding tube had been placed and antibiotics started. Social services had been consulted to arrange nursing home placement.

I suppose, given our history, I could have left it at that. But I felt somehow responsible for him, since I was the one who had contacted him, opening his life ten years earlier. So I caught a plane out to Kansas City to see for myself what was going on. When I walked into his room, I was shocked to see how very thin and fragile he appeared—a shadow of his former self. He now weighed no more than ninety-five pounds, and I realized that he had been deteriorating for some time and hadn't told me.

Despite his condition, he appeared to be genuinely happy that I had come. Motioning to a chair next to his bedside, he tried to speak, but he was very difficult to understand. His voice was barely audible as a result of the ravages of Parkinson's disease. He had not been able to swallow for quite a while and had been hoping he would just die of natural causes at home. A feeding tube was the last thing he would have wanted. He spent what little energy he had that day inquiring about me, my husband, and his favorite topic, Rory. He was particularly interested in hearing all about Rory's Little League baseball adventures, reminding me that he had loved to play baseball as a child himself.

Finally exhausted, he took my hand and asked if I remembered the conversation we once had about how he wished to end his life with dignity if he were ever in this situation. When I said that I did, he replied, "Then you know what to do. Please help me now."

Without further discussion, he changed the subject again before drifting off to sleep.

I knew that I needed to do this one last thing for him. I asked to speak with his attending physician, explaining that I was a physician myself and that he was in his right mind. They were to declare him a DNR (Do Not Resuscitate), discontinue use of the feeding tube, stop food and hydration.

I knew that without hydration it would take about three days for him to slip away into a painless coma. They would also give him morphine to ensure his comfort. Explaining all of this to him, I again inquired if he was certain this was what he wanted. I could tell by the look in his eyes that he had no doubts.

I really don't know what I was hoping for as I slipped back into his room in the middle of the night to say one last goodbye before catching my plane. When I decided to seek him out ten years earlier, I really just wanted him to tell me that he was sorry. Sorry for abandoning his family of four children when I was only four years old; sorry for never sending any child support, a Christmas or birthday card; sorry for abandoning us without caring for what would become of us, for depriving us of his love. When he saw me standing by his bed in the sterile moonlit room that night, waiting to exchange our last goodbye, he motioned me down, threw his arms around me, and whispered into my ear, "I love you."

The call came forty-eight hours later. My father had slipped into a coma and died peacefully at sunrise. Even though it was expected, when I heard the news, it felt like a knife piercing my heart, exactly as it had when I saw him for the last time when I was just a little girl. So much lost . . . so little said . . . such a sad man who lived such a sad life. He has left me again for the final time. His loss . . . my sorrow.

You left nothing of yourself for me to mourn, almost as if you had never existed. I know you wanted it that way. I also know deep down that you were sorry, even if you never said it out loud to me.

CHAPTER 24
Prancer Tales

There are situations, despite all we try to do, when a diagnosis can be missed. For a diagnosis begins as a suspicion, shaped by the patient's narrative and followed by piecing together a puzzle from the fragments of complaints, testing, and silent clues. Sometimes, a little intuition goes a long way. That and the help of a very intuitive therapy dog.

Prancer, a gentle creature with a coat the color of autumn leaves and eyes as deep and soulful as the ocean, was more than just a golden retriever. She was a healer, a silent confidante, and a source of unspoken comfort. She was my dog, trained and certified through the Pet Partner program to accompany me to comfort patients in my clinic.

On one occasion, I was asked by a colleague to consult with Kaya, a pleasant forty-year-old woman from Kenya. She had been too ill to function, ever since she immigrated to the United States a few years ago, and no one had been able to sort this out.

"She has so many symptoms," my colleague stated, "it's hard to pinpoint anything definitely neurological. But I love to send these types of patients to you because neurologists are the expert history takers. Please see her, see what you think."

Kaya was a strikingly beautiful woman, stylishly adorned in a bright orange-and-green cotton dress with cornrow braids adorning her shoulders. She seemed timid at first.

"I don't know why I sent here," she chimed in broken English as she sat before me. "I already see every doctor, have every test."

I had set aside an hour for our consultation. I tried to gain her trust by exchanging pleasantries about her native Africa, which I had recently visited. Then I suggested introducing her to Prancer, who had an aura of serenity about her that soothed most troubled hearts. But Kaya was having none of it.

"Oh no, please don't let her near me," she pleaded. "I don't like dogs."

Prancer always waited until I was in an exam room with a patient before gently scratching the door to ask for entry. When I didn't let her in, she would just lie down quietly outside the door. This was one of those days.

Kaya had come prepared with a two-inch spiral-bound book, complete with color-coded tabs for all her specialty visits and prior testing. Green was nutrition, purple rheumatology, red cardiology, blue physiatry (rehabilitative medicine), yellow primary care, and white was naturopathy. The only specialist she had yet to see was psychiatry, which was next on her list, after me.

"In Kenya I was forced to experience much physical and sexual abuse as child," she said with broken English. "They try to tell me my symptoms are all in my head, but I cannot accept that."

Kaya had many somatic complaints: total body pain, inability to sleep, irritability, tingling sensations all over, general malaise, rashes that would come and go, mouth blisters, dizziness, and so on. Her examination failed to reveal a neurological deficit. And review of all her extensive testing, which included MRIs, both brain and spine, vials and vials of blood work, an overnight sleep study, and extensive cardiac testing, all failed to shed any light on her array of symptoms.

At last, after a detailed history, examination, and review of her prior testing, I agreed that perhaps given her strong history of childhood abuse, a psychiatric consult would be appropriate. As I walked her out of the exam room, she finally caught a glimpse of Prancer, patiently waiting outside the door.

"Oh my!" she declared. "She is very beautiful and sweet. I'm sorry I didn't let her in. May I pet her now?"

Prancer usually didn't take well to being denied entry, and I fully expected her to just get up and walk away, but she allowed Kaya to pet her soft golden fur, drawing comfort from her wellspring of unconditional love. Then, to my surprise, instead of coming back to the office with me, Prancer padded after Kaya to the reception area with her tail wagging rhythmically. When she failed to come when I called her, I walked out to see what the matter was. I found her curled up under Kaya's chair and she refused to come out.

This behavior was so unusual for Prancer that I decided she was trying to tell me something. So I went out and told Kaya that I had changed my mind after all. Even though her testing had been quite complete, I thought it might be worthwhile to repeat some of the autoimmune testing the rheumatologist had performed six months ago. Perhaps another look again now might be more revealing.

As it turned out, Kaya now had critical markers of autoimmune disease—lupus, an elusive and unpredictable disease, which is often disguised in mystery and misunderstanding. To the sufferer, lupus can be a constant, unseen companion. It is a thief of vitality, stealing away energy and replacing it with an ache that resonates in every joint, a fatigue that seeps into every muscle, turning even the simplest endeavors into monumental tasks. In moments of remission, lupus retreats into the shadows, offering a deceptive reprieve. Yet, like the changing tides, lupus often returns with a vengeance, its flare-ups unpredictable and fierce. Lupus weaves its way through the body, touching organs with malevolence. The heart, the kidneys, the lungs, and the brain are all potential victims of its wrath. It is a master of disguise, mimicking other illnesses, making its diagnosis a puzzle that requires patience and persistence to solve. The challenge is often in the diagnosis. With proper treatment, most patients go on to live normal lives.

This was indeed a happy ending. At last, an answer for this patient, albeit not a neurological one, but one that I almost missed. I referred Kaya back to her rheumatologist for treatment.

If it were not for Prancer's intuitive superpower, sensing just how sick and sorrowful Kaya was that day, I might never have repeated her blood work. I was reminded again just how incredible the healing power of Prancer's intuition could be. Words were unnecessary with Prancer, as her empathy transcended language, reaching into the depths of human sorrow, offering her patients a glimmer of hope.

Thank you, Prancer!

CHAPTER 25

Angels Among Us

I believe in angels. Let me get that out there right now. I believe they exist to help us when we need them, and I suspect some of them roam the earth in human form, sent here to shed some light on our world and inspire us by example.

Angels make me think of Andrea.

I first met Andrea when she sought a consultation for poor balance. At the age of just twenty-six, my initial suspicion was a condition such as multiple sclerosis, which is more likely to affect the young adult. I became concerned during the interview, however, when she described her symptoms: progressive gait ataxia (walking like she was drunk), vertigo, nausea, occasional vomiting, and headache. Her graceful stride had become hesitant. Her hands, once steady, now trembled with a persistent quiver. Her symptoms had been progressive over several months, and yet she postponed seeking help, convincing herself that her perpetual dieting was responsible for these symptoms.

"It feels more and more like my body is no longer obeying my will," she said. "It's almost as if my mind was inhabiting someone else's body."

Andrea went on to explain that she had been on the heavy side most of her life, and only recently had she been successful at dropping some of her weight. She wanted to marry and have a family someday, but she had been shy about meeting men because she was so self-conscious. She had recently landed a job that paid well enough for her to get out on her own, and she was anxious to begin a more active social life.

Andrea had a beautiful aura about her. She was tall with shoulder-length straight blonde hair, green eyes, scattered freckles, and her demeanor was open and friendly. I tried to dismiss the thought that bad things only seem to happen to nice people.

It was perhaps the persistent and progressive nature of her headache that concerned me the most. This was not typical of MS.

Andrea's examination was concerning for dysfunction that localized to the cerebellum, which is the balance center of the brain. Hence the gait ataxia, vertigo, and nausea.

Though it does not command the intellect or emotions, the cerebellum's influence is profound. It is the custodian of balance and harmony, turning clumsy impulses into smooth, purposeful motions. Nestled at the base of the brain, the cerebellum functions as the silent conductor of the brain's symphony, orchestrating movement with an elegance that belies its modest size. It is like a masterful puppeteer, deftly pulling the strings to ensure that every muscle, joint, and limb perform with precision and grace. It is responsible for the fluidity of a dancer's leap, the surety of a surgeon's hand, and the effortless coordination of a child's first steps.

Andrea's brain MRI confirmed something much more sinister than multiple sclerosis. Hidden within the sanctuary of her cerebellum, a malignant tumor had begun an insidious invasion. The center of her cerebellum was swollen with edema and lit up with the gadolinium dye–enhanced images—a sure sign of a malignancy.

Wasting no time, I called Andrea in to discuss the results, and suggested she bring along a family member. To my surprise, she showed up alone.

"Wouldn't you like your parents with you to hear what I have to say?" I asked.

Andrea bravely answered,"I am all grown up and can handle whatever you have to tell me. Besides, it is just me and my mom. My dad left when I was small, and I haven't heard from him since. If the news is bad, it will just devastate my mom. I want to shield her from worry as long as I possibly can."

As the tumor's shadow loomed large, casting its pall over her cerebellum, Andrea embarked on a journey into uncharted territory. An excellent team consisting of neuro-oncology, neurosurgery, and radiation specialists descended upon her with the unified goal of saving her life. Her biopsy revealed a malignant tumor, an anaplastic astrocytoma, which was grade III out of IV. At her age, her five-year survival rate was estimated to be about 58 percent treated aggressively with surgery, radiation, and chemotherapy.

I saw Andrea infrequently when she was undergoing treatment. True to her character, she remained strong and optimistic with the heart of a warrior, intent on reclaiming the rhythm of her life, step by awkward step. When she was returned to my care for monitoring, we were both hopeful. Her treatment had gone well, with no tumor recurrence over ten months. She was working with physical therapy daily, intent on reclaiming control of the balance and dexterity that her relentless adversary had stolen from her. She underwent follow-up brain MRIs every three months with me and checked in with her neuro-oncologist every six months.

Andrea was finally back at work and excited to tell me that she had met a nice young man. She still struggled to perform a tightrope walk, but for all practical purposes, she had regained her dexterity. At first, we both met with trepidation on the day she had her follow-up brain MRIs, fearing for what we might find, but with each stable study, we became more and more assured. I was always happy to see her on my schedule as she seemed to radiate joy, which was so welcome at the end of a long day.

It had been about a year when Andrea showed up at the end of my day so we could review her latest MRIs together. She had no new complaints, but I wonder now if she would have told me if she did. Her brain scan appeared unchanged to me, and her examination remained stable. As she was telling me all about her life and the new love she had found, however, my eyes happened to glance down on the scan, below the cerebellum and below the brain stem, where I could barely visualize the top of the spinal cord. Something wasn't right there. I saw some leakage of dye, what we call enhancement, suggesting inflammation.

I was hopeful that this was just a distortion of the image due to technical factors, as the cord was not directly imaged.

Dropped metastasis to the spinal cord from a brain tumor is extremely rare. Not wanting to alarm Andrea, I ordered a full spine series and told her it was just to assuage my cautious nature.

The news was not good. Even though Andrea's brain MRI showed no recurrence of tumor, she had extensive metastasis to her spinal cord. Unfortunately, anaplastic astrocytomas, such as Andrea had, have been known to become more aggressive after treatment. Her prognosis for survival was now two to three months.

This brave warrior had prepared herself for the news before I even called her in. Andrea didn't scream about how unfair this was or even cry in front of me. All she could think about was her mother.

"How am I ever going to tell her? I'm all she has," she said.

I knew then that the incredible spirit of this young woman would find a way to somehow soften the news. She was such a bright light in this world, undaunted by the tragedy thrust upon her at the very beginning of her life.

Andrea instinctively knew that the life we are given is such a fleeting gift, a gift to be cherished, for we never know how long it will last.

CHAPTER 26

Please Don't Cry

Letter to My Brother

Blaine, you were my best friend growing up. Unable to pronounce my name when I was born, you affectionately called me Susie. I remember when I was four and you were six, you decided we were going to build a tree house. We gathered the necessary tools—sticks, hammer, and nails—and set off on our adventure. Though I tried as hard as I could to hold the nail steady while you hammered, I suddenly felt a rush of pain as the hammer came down on my thumb. As I was about to wail, you were there to comfort me and assured me that if I held back my tears, it wouldn't hurt as much. I believed you, my brother, and held back my tears that day for you.

Fifty-some years later, I was finishing up the day at my neurology clinic when your daughter Brittany called. "Dad's had a heart attack," she said. "Please come." The week before, I was so delighted when, after not speaking with you in a while, you happened to call just to tell me you loved me. Since my move three thousand miles away fifteen years earlier, we rarely saw each other. I wonder now if you had some premonition about what was about to happen that day.

Assuring Brittany I was on my way, I caught the red-eye to the East Coast. It was unimaginable to me that you could possibly have had a life-threatening cardiac event. You were too young and had no prior history of a problem. Growing up, you delighted so in playing practical jokes on me that, in my denial that day, I tried to imagine this might just be another one, conjured up to finally get me to come out and visit. Do you remember when I was at Notre Dame, and you surprised me with a visit one football weekend? You brought an old high school boyfriend of mine, knowing he was the last person I ever wanted to see, and we laughed the whole time! You were always able to make me laugh.

This time, upon landing, I went right to the hospital and was told you were in the ICU. This sacred place, where life and death are decided, with its antiseptic smells and familiar sounds of the machines, was all too familiar to me. Only this time I was not an attending neurologist but a visitor, here to see one of my own, whom I have loved dearly for all my life.

As I walked into the ICU, I immediately knew which room was yours, as it was surrounded by your entire family, all of whom appeared anxious but relieved.

"Dad just came back from the catheterization and had stents placed in his coronary arteries. He's already looking better," Brittany said as she stepped aside so I could approach your bed. I expected to see you sitting up, joking around as usual.

I was immediately alarmed. "Why is Blaine on a ventilator? Is he sedated?"

Your wife, Annie, was too distraught to speak, so your daughters, Brittany and Allyson, filled me in about the night's events: how you had awoken in the middle of the night unable to breathe and saying you were dying, how you were rushed to the nearest hospital where you underwent emergency cardiac catheterization. The procedure was unusually long—four hours. The cardiologist had just left, informing your family that you were now stable.

You did not look stable to me.

I first checked your IV bags and noted that you were not on any sedating medications, so you should be awake by now. I called your name, pinched your nail beds, trying to elicit a response to the pain. There was no spontaneous movement or withdrawal from pain. I manually opened your eyes; your pupils were pinpoint and nonreactive. There were no eye movements when I turned your head from side to side, a clear sign of extensive damage to the brain stem, which is the control center for heartbeat and respiration. Finally, when I scratched the bottom of your feet, I saw your toes turn upward reflexively, a positive Babinski sign indicating global brain dysfunction.

It was clear that you had suffered a massive stroke during the catheterization, yet the hospital staff seemed to be either unaware of it or they were covering it up.

Not wanting to alarm your sweet wife, Annie, or Allyson and Brittany, I told your nurse that I wanted a neurology consult right away. The cardiologist, a middle-aged man with a stubby white beard decorating his weary expression, then appeared at the door.

"I really don't feel that a neurology consult is necessary," he said. "Blaine is just slow to wake up from the sedation. This was one of the most difficult catheterizations I have ever had to do. Your brother's vessels were so thick with calcifications from extensive vascular disease that I had poor venous access. After hours of trying, I was finally successful in placing stents in two coronary arteries."

"With all due respect, I am a neurologist and there definitely is a problem," I replied, frightened about what I knew was happening.

"Well, since it's Friday, let's give him the weekend and see how he is on Monday," the weary cardiologist answered.

Now feeling even more anxious, I said, "That's not good enough. His exam indicates there's been massive brain injury. I must insist on a formal neurology consult and an MRI today."

Realizing I wasn't going to accept his excuses, he called a Friday afternoon neurology consult.

This scenario reminded me of something from my days as a neurology resident. I recall seeing a patient in consultation for stroke symptoms noticed immediately after a cardiac catheterization. The cardiology resident had called the consult and I responded right away. In the middle of my examination, the attending cardiologist who had performed the procedure walked in.

"What is she doing here?" he barked. "I don't want a neurology consultation. Get her out of here!"

The consult request was canceled, erased from the chart like it never happened. The implication being that a complication during the procedure itself would not look good for the doctor's statistics, especially in this highly competitive academic center.

The neurology consultant on this day, a young man with an East Indian accent, was not much more helpful than the cardiologist. "I plan to give Blaine the weekend to see how much he improves," he said. "If he's no better on Monday, we'll consider a brain scan then."

As it was you, my brother, lying lifeless on that bed, I insisted that an MRI be performed right away. The MRI finally ordered, we never saw the neurology consultant again. The message delivered through the nursing staff was that I could read the MRI myself. The consultant had gone home for the weekend.

Unbelievable!

Now gathered with your entire family—Annie, your daughters, stepsons, all their spouses and our two other siblings, Judy and Geoff—we waited for the MRI results, hopeful that my preliminary assessment was wrong, that there would not be an extensive area of damage as I initially anticipated.

Finally, one of the nurses slowly wheeled into the room a computer terminal on which your brain images were stored for my review. The results were devastating. Both cerebral hemispheres and the entire brain stem showed recent infarction or death of tissue, sparing only one small area in the speech center. In all the years I had practiced neurology, I had never witnessed a stroke this extensive!

The heart attack was the inciting event responsible for a blood clot, but then either a clot or vascular debris (calcium, cholesterol) must have dislodged somehow during the difficult catheterization procedure, showering into all the major cerebral vessels supplying multiple parts of the brain—something we call multiple embolization.

At this moment I had to summon up my neurology training. I could not allow myself to fall apart and watch your family—our family—suffer through the agonizing next several days that were about to occur without intervention. With such extensive death of brain tissue, there would be massive swelling, likely causing the brain to herniate or compress downward into the small spinal canal—a fatal event. Sadly, with no hope of any meaningful neurological recovery, you should at least be allowed to die with more dignity than this.

With the neurology consultant AWOL and the cardiologist clueless, I was put in the position of telling your family—our family—what had happened to their husband and father, and how utterly dismal the prognosis was. In my professional capacity, I have performed this sad task many times before, but never for my own family. My words seemed to take a while to sink in as Annie and your daughters asked their questions repeatedly. Annie, unwell herself, a result of pulmonary sarcoidosis requiring oxygen supplementation, just kept repeating the events of the evening, unable to comprehend the finality of all of this as she kept insisting that you had been fine just yesterday. There had never been any warning signs that you had a heart condition. At least nothing you ever mentioned to her, which was so like you, as Annie's health was always your priority. As neurologists we often try to soften the picture by using the term "no meaningful recovery," but this was worse—far worse.

There was no time to wait and see, no time to allow for gradual acceptance of the inevitability of the situation, as brain death would certainly occur over the next twenty-four to forty-eight hours as the brain started to swell. I advised withdrawing life support sooner, rather than waiting for this more devastating event.

All in agreement, your daughters called in a priest for the last rites, something they felt you would have wanted with your Catholic upbringing. One by one, each family member took their time by your bedside, saying goodbye. The atmosphere was heavy as we gathered around your bedside one last time before withdrawing life support. Our oldest brother, Geoff, broke the silence with some comic relief: "Before you go, Blaine, I have a confession to make. I voted for Hillary."

Laughter ensued for a few minutes, something you no doubt would have enjoyed.

The ventilator now silenced, the tube down your throat removed, you slipped away peacefully, surrounded by all those you loved most in the world. Only forty-eight hours before, you had been out to dinner with Annie after a routine day of work.

At your funeral, I seemed to be the only one unable to cry. I simply couldn't find a way to shift back out of the professional mode of doctor so that I could become just your sister during that time. My hard-learned skill of detaching myself from the situation initially helped me cope with the devastating task at hand, but it now prevented me from grieving as I should. I will never forgive the way those doctors mismanaged your case, making it necessary for your sister to have to get so involved.

That was a day I wished I wasn't a neurologist. But for our family's sake, I'm glad that I was, and that I was there for them and for you.

It hurts knowing that you are no longer part of my world.

Holding back my tears this time for you, Blaine, is not making the pain of losing you any more bearable. Just thought you would want to know.

Susie

CHAPTER 27

It Wasn't the MS

This is a story about Tim, a young man who had already faced more than most. Diagnosed with multiple sclerosis at an early age, he battled through the physical and emotional toll of the illness with remarkable strength, determined not to let it define him. Against the odds, he rose above his condition, showing a resilience that inspired everyone around him. But in his fight against the chronic pain that came with his illness, a new enemy emerged—one far more insidious. Prescribed opiates to manage his pain, and benzodiazepines to manage his spasticity, he gradually became addicted, and what began as relief from his suffering eventually spiraled into dependency. In the end, the same drugs meant to ease his burden took his life, leaving behind a heartbreaking story of strength, struggle, and a tragic end.

The last time I saw Tim, a thirty-nine-year-old man I had treated for MS (multiple sclerosis) for over twenty years, he was doing extremely well. A triathlete, he had been stricken with MS at the age of nineteen, requiring extensive rehab after a particularly severe spinal cord attack resulting in paralysis and a neurogenic bladder, with loss of bladder control. The treatment for MS has come a very long way over the past twenty years, allowing for more and more aggressive treatment that has the potential to significantly slow down and, in some cases, arrest the progression of disease. Tim, anxious to do everything he could to preserve his hard-earned return of full function, save for residual bladder symptoms and mild spasticity, took advantage of all that medical science had to offer. He was even able to begin training to cycle competitively again.

Now married and actively employed with his own bicycle repair shop, our visits were mostly spent reviewing his latest very stable brain and spine MRIs, managing his disease-modifying therapy for his MS, and rejoicing in all his accomplishments and the happiness that he had found in life. He and his wife, Amy, were even talking about trying to have a baby, something I thought would never happen given the depth of his despair in the early days of his diagnosis.

Early in the course of his disease, during his rehabilitation from his severe spinal cord attack, a well-meaning rehab specialist/physiatrist started Tim on narcotics for pain and later benzodiazepines—Valium-like drugs to help with his spasticity complaints. His use of this crutch quickly spiraled out of control, such that I was alarmed at the doses he was taking on my very first visit with him. I cautioned him about the need to taper off these substances. But suspecting that I did not approve, Tim made sure to prevent me from getting involved in this aspect of his care, insisting that he would be unable to function without them. Over the years, I saw his use of these drugs wax and wane. He would attempt a taper under the direction of his treating physiatrist, but somehow he would always develop some malady requiring escalation, despite optimum control of his MS.

I was secretly delighted when the rehab specialist managing Tim's pain and spasticity retired, as it allowed me to refer Tim to a highly respected addiction specialist who had been very helpful to me with other patients. Tim, always engaging and likable, hit it off with Dr. G. right away. I was very pleased with Tim's progress over the years under his guidance. But he still claimed he needed some assistance with his pain and spasticity management. It was clear Tim had a true addiction driving his need for these drugs, but he would never admit to this, insisting he needed these medications for his residual MS symptoms.

It had been more than six months since I had seen Tim in my office, so I was pleased when I saw him on my schedule that day. Accompanied by Amy, he informed me that he had just been discharged from a detoxification unit. He was off all narcotics and benzodiazepines. It had been arranged by Dr. G., who finally laid down the law: detox or discharge from his care. Tim seemed extremely happy, pleased that he was finally off all addictive medications and ready to start life anew—perhaps start that family they had wanted. Amy, beaming, just wanted to thank me for all the MS care I had offered Tim over the years that had kept him so stable. Tim's neurological exam was normal, so I planned for his follow-up yearly brain and spinal cord MRIs and, bidding him adieu, told him I would call with the results. "See you in a few months, Doc!" he said as he walked out of my office that day.

That was the last time we talked. The following day, I received a tearful call from Amy. Tim had gone out to his shop to work on some bicycles, something he often did late at night. When she awoke and found that he had never returned to their bed, she went looking for him. She found him slumped over in his chair, cold, gray, and unresponsive! Next to his lifeless body was an unmarked bottle of pills, all gone except for one lone OxyContin. She had suspected for some time that he kept a private stash somewhere, which explained why she would sometimes find him asleep and drooling in the middle of the day.

Addiction is a powerful foe. Just when you think you have conquered him, that incredible craving sneaks back in, forcing you to take one last one—just to help you sleep, perhaps, or to get through a difficult meeting. It is as if there is a little red demon perched on your shoulder whispering in your ear. The better you seem to be doing in life, the louder the whisper: "Just one more—what can it hurt?" At least that's what you convince yourself.

The patient recently detoxed is particularly vulnerable. Not having used the drug for a while, their tolerance is much lower than it was when they were using frequently. But they often don't recalibrate their dosage accordingly. Instead, they take as much as they were using when they left off. Now, just a fraction of that might be enough to fatally depress the respiratory center.

While Tim survived and conquered the demon of MS that plagued his young life, his real demon, unknown to him at the time, was the drugs he abused as an excuse for his symptoms that mercilessly took his life in the end.

Tim's story is one of both triumph and tragedy. He fought a courageous battle against multiple sclerosis, showing a strength that few could match. But in his pursuit of relief from his pain, he found himself caught in the grip of addiction—a battle he wasn't prepared to fight. His death from an overdose is a heartbreaking reminder of the fine line between treatment and dependence, and the devastating consequences that can arise when pain management turns into something far more dangerous. Though his life was cut short, his resilience in the face of the MS remains an enduring testament to his spirit. And perhaps his story serves as a call for greater awareness of the risks that come with managing chronic pain, so that others might not follow the same path.

CHAPTER 28

Lightning Lessons

Growing up, I loved thunder and lightning storms. I remember the thunder shaking the walls of the house as the heavens unleashed their fury upon the earth, followed by the ethereal dance of lightning igniting the sky.

As kids, we would all hunker down in the safety of our home as the storm approached, the dark, billowing clouds gathering, foreshadowing the unleashing of raw power. The air would be heavy with anticipation, crackling with electricity, the scent of ozone permeating the atmosphere, warning of the spectacle about to occur.

Even now, when I hear thunder, I'm reminded of the beauty of the Northeast, the magic of those storms, and the way they used to make me feel. Since moving to the Pacific Northwest, where lightning storms are rare because of the cooling temperatures off the Pacific, I find myself longing for that feeling, but I have found solace in the peacefulness of the rainy days and the hopeful feelings conjured up from the fresh, balsam-scented air afterward.

And the welcome absence of earthbound lightning bolts.

The odds of being struck by lightning are approximately 1 in 1.9 million. Imagine for a moment what it must be like to be struck by lightning. That moment when the mortal and celestial realms intermingle as a celestial javelin plummets from the sky seeking an earthly connection. With a single, blinding flash of incandescent brilliance, the lightning finds its target—an ill-fated victim, unsuspecting and unwittingly positioned in the crosshairs of nature's indiscriminate wrath. The bolt, with its searing intensity, becomes a prelude to transformation, as all that energy is discharged into the victim in a split second, coursing through the victim's mortal body, with an otherworldly precision. The thunderous boom that follows testifying to the event.

Then, as abruptly as it arrived, the lightning retreats, leaving behind a transformed victim. Charred remnants of fabric, singed hair, and profound numbness and shock initially invade the senses for those lucky enough to survive the attack. Others are not so lucky, as the electrical charge frequently stops the heart. For those who survive lightning strikes, the neurologic complications can be devastating.

Survivors often describe the sensation as sudden and overwhelming, almost beyond comprehension. Many recount a blinding flash of light, followed by an intense, burning heat that feels as though it's radiating through their entire body. The pain is often described as excruciating, a searing agony like their nerves are on fire, accompanied by the violent force of the strike, which can feel like being hit by a powerful shock wave or a massive jolt of electricity. Some report feeling temporarily paralyzed or numb, unable to move or even breathe as the current courses through them. Others describe a tingling sensation, often in the extremities, or an intense buzzing in their ears, as though they are vibrating. Many lose consciousness or have memory gaps, waking up disoriented, with burns, torn clothing, or, in some cases, thrown meters from where they were originally standing.

Emotionally, survivors often speak of confusion and fear, unsure of what just happened, followed by a deep sense of shock and disbelief that they survived such a life-threatening event.

Residual symptoms from a lightning strike can range from temporary dizziness, confusion, and difficulty concentrating to more serious long-term medical complications, including paralysis, cognitive impairments, and limb weakness. It goes without saying that lightning strike survivors may also experience psychological and emotional complications, such as post-traumatic stress disorder, depression, and anxiety. Other common issues include chronic pain, physical impairments, and sleep disturbances.

Practicing neurology in the Northeast, I had the occasional rare opportunity to see a lightning survivor. They usually presented with a bizarre array of peripheral nerve injuries where the lightning bolt entered and exited their body.

One such patient, John, was a rugged-looking man in his mid-thirties who had been struck by lightning when he was helping a friend put a metal roof on his house. He was incredibly lucky, in that the lightning bolt had only singed his hair and left a small scar on his arm where the lightning entered. However, once it entered John's body, it gained access to the peripheral nerves communicating all the way to his spinal cord. The nerve injury was quite severe. He had numbness in his hands and feet, along with severe pain that radiated from his shoulder to his fingertips.

After a few months of occupational therapy, John was able to regain some of his strength and dexterity, but he still had residual nerve damage that prevented him from doing certain activities. Again, he was one of the lucky ones.

Then there was Arthur, a middle-aged man who had been working on a power line when the lightning struck the pole he was on, causing him to fall to the ground. Fortunately for him, a passerby knew CPR and, recognizing that Arthur's heart had stopped, resuscitated him on the scene, saving his life. I saw Arthur to evaluate a peripheral nerve injury that had occurred where the lightning bolt had coursed through his leg. I needed to test the conduction of the nerve, which is done using small shocks. I was concerned, however, that he would be unable to tolerate the testing because of PTSD from the shock. I was surprised that he barely felt it, having no recollection of the lightning event whatsoever.

I never expected to see a lightning injury in Seattle, where we rarely have a lightning storm. But to my surprise, we had an incredible thunder and lightning storm one evening that even made the nightly news. A few months later, Joshua, a freelance writer in his early thirties, came to see me in consultation. Dressed casually in jeans and a white button-down shirt, he smiled, revealing numerous missing teeth.

"Hey, do you remember that cool storm we had a few months back?" he said.

"I certainly do," I replied.

"Well," he said, "it was so cool that I felt I had to stand out on my balcony to watch the whole thing. So there I was, leaning on my metal railing, when I saw this blinding flash of lightning hit the tree ten feet in front of me. And that's when the lights went out! When I woke up, I was twenty feet inside my living room on the floor, confused, and all the fillings had been knocked out of my teeth!"

The electricity had been drawn to his moist fillings!

Joshua had been home alone, so it's unclear how long he had been unconscious. He had no memory of getting hit. It's quite common to have a short period of retrograde amnesia after any type of concussive head injury. And since the lightning catapulted him backward twenty feet, crashing his head on the floor with such force that he lost consciousness, that certainly qualified.

Joshua was very lucky to have survived this lightning strike, and he initially thought that his only problem was going to be the cost of having all his dental work replaced.

I wondered to myself, but dared not ask after the fact, if he was really that uneducated about the danger of standing out in the rain holding on to a metal balcony with lightning all around, or if his judgment that evening was perhaps altered by a substance.

As it turned out, Joshua's main difficulty was that he was unable to resume work as a freelance writer because of cognitive dysfunction, emotional instability, and insomnia.

As a human being, he emerged from this experience with a newfound reverence for the interplay of forces beyond human comprehension. His residual cognitive deficits became a living testament to the ephemeral dance between the celestial and the mortal, bound by that electric embrace that forever altered his existence.

My guess is that is the last time Joshua will stand on his balcony during a thunder and lightning storm.

CHAPTER 29

Medical Gaslighting

There's a little secret in health care that hasn't gotten the attention it deserves. Medical gaslighting in women.

Gaslighting occurs where medical professionals overtly dismiss symptoms, leaving crucial health issues untreated, causing even the patient to doubt themselves. The patient's lived experience, as they describe it, becomes somehow questioned, minimized, or outright dismissed. It can be as subtle as a smirk, a raised eyebrow, a sigh of impatience that punctuates their narrative, as if to imply that the patient's genuine suffering is an exaggeration. The doctor's words, once meant to heal and soothe, can instead become sharp instruments, cutting away at the patient's confidence and sense of reality. When this occurs, there is often a clear power imbalance at play.

Gaslit patients are more often women, people of color, older adults, and those suffering from obesity. Studies have shown that this has led to women, on average, being diagnosed with diseases four years later than men. Whether it is heart disease labeled as anxiety, an autoimmune disorder attributed to depression, or ovarian cysts declared as "normal period pain," many women's health issues are more likely to be misdiagnosed or dismissed by doctors as something less serious. Even women doctors are not immune to this phenomenon, as I experienced it myself.

I have never been one to seek out medical care, as I've been fortunate to have been in excellent health most of my life. So, when I presented to my GYN for my yearly follow-up, I brought up my first health complaint in the twenty years I had been seeing her—abnormal bleeding. We were colleagues and she knew I was not one to complain, so I'm happy to say that I was taken seriously. She scheduled me right away for an in-office uterine biopsy. Thankfully, it was normal, so I thought no more of it. At least not until the symptoms recurred six months later. Back I went, and this time I underwent another painful biopsy and a pelvic ultrasound.

"I'm pleased to tell you," she said, "that the biopsy results were normal again and there was nothing alarming on your ultrasound, just a benign-looking polyp."

Feeling much reassured, I retreated to work and put it out of my mind.

Six months later, once again symptomatic, I questioned myself before returning. Maybe this was just another normal inconvenience of aging. I was due for my yearly follow-up and I knew my trusteed colleague was soon retiring, so back I went and let her know I was symptomatic again. She was hurried, getting ready to close her practice, with plans to leave for an extensive and long-awaited tour of Europe with her husband.

"I'm not doing another biopsy," she said. "There's nothing wrong with you. This is all just stress. I will provide you with a few names of gynecologists in the area for you to follow up with after I leave, but you don't need to be seen for a year."

Wishing her well in her retirement and thanking her for all her care over the years, I left. Only this time I didn't feel so reassured. At this time of my life, I was no more stressed than I'd ever been, and I resented her dismissive evaluation that nothing was wrong with me. A nagging whisper in my head told me I needed to get another opinion, and soon.

My new GYN became alarmed when she saw my prior ultrasound. "You should have undergone a hysteroscopy [scraping of the uterine lining with biopsy] and removal of that polyp a year ago," she said. "I'll get you on the schedule right away."

After the procedure, I felt reassured when I didn't hear from her, as an abnormal biopsy would have been flagged and I surely would have gotten a call. When I showed up for my follow-up with her two weeks later, I relayed a particularly vivid dream I had the night before. I had dreamt that I was sitting in her office, and she was telling me I had a malignancy. We both laughed as she quizzically looked at her computer terminal.

"Wait a minute," she said. "I don't see your results here. Let me check with my assistant." A few moments later, as she glanced at the paper results, a look of dread came over her face. "I am so sorry," she said. "I don't know how this could have been missed. Your biopsy did reveal a malignancy. This should have been brought to my attention immediately."

I was in the OR within a week for a total hysterectomy. Knowing how long it took to get this far and just how critical it was to diagnose uterine cancer early, I was terrified. The prognosis for endometrial cancer is generally good if caught early, before the cancer has spread through the uterine wall. In these cases, 95 percent of women are still alive after five years. When the diagnosis is delayed, as in my case, and the cancer has invaded through the uterine wall, only 25 percent of women are still alive after five years.

Only by the grace of God, my cancer was still Stage I. There was no spread. Had I followed the advice of my now retired GYN and not listened to my own fear, I probably wouldn't be here today.

And then there was Mia.

Mia traveled three hours to see me. She had sought me out because I was a woman, and based on my online reviews, she was hopeful that I would listen to her and have more empathy than the doctors she had previously encountered. She was an attractive, somewhat timid twenty-eight-year-old woman, accompanied by her husband, with numerous medical records in hand. My heart went out to this young woman who was not only morbidly obese but confined to a wheelchair. She had a very sad story to relate.

"I didn't always look like this," she began. "I've been overweight for most of my life, despite desperately trying to lose weight. Several years ago, I went on a strict diet and managed to lose eighty pounds over eighteen months, just in time to fit into my wedding gown. Shortly after our honeymoon, I started to develop severe low back pain. I saw numerous doctors, my primary care doctor, orthopedic specialists, and a physical therapist, but no one could figure out what was wrong with me. Then I developed severe lower abdominal pain and was admitted to the hospital. Once again, after many tests, the doctors were at a loss about what was wrong with me. I was discharged to go home with no diagnosis and no medication or follow-up. I was starting to doubt my own sanity, my own truth. I didn't know where to turn and then the weakness began."

Mia tearfully went on detailing her entire story as her husband held her hand. At first, she had trouble climbing the stairs and then she started falling. Then she noticed she was even having difficulty lifting her arms above her head to fix her hair.

"I was becoming frightened, so I returned to the ER," she said, "where I was well known by now. I waited in the waiting room for hours before they took me back. After a cursory exam and MRIs of my spine, I was told that nothing was wrong with me."

Mia's weakness progressed by the hour, until she described being unable to walk out of the ER. She was finally admitted to the hospital for observation. Within twenty-four hours, she started feeling pins and needles in her extremities, and then numbness progressing up her legs.

"By this time," she said, "I was becoming hysterical, as no one seemed to be taking me seriously."

Mia remained hospitalized for nearly two weeks until they finally transferred her to a rehab facility for physical therapy, still without a diagnosis. "I was now completely wheelchair bound and convinced it must be all in my head."

It wasn't until Mia was discharged from there that she got in to see a neurologist for further testing. With electrical nerve testing, he was finally able to confirm that she had developed a progressive peripheral neuropathy. He initially tried treating Mia as an outpatient with a course of steroids, but she never improved.

"The devastating part of this was that I put back on all the weight I had lost before our wedding, and more," Mia said. "All I want to do is cry. Please believe me."

I reviewed Mia's extensive records. She was finally diagnosed with chronic inflammatory demyelinating polyneuropathy (CIDP). This is a rare condition that comes on insidiously and progresses over weeks to months. It can be heralded with abdominal pain and commonly also back pain. As the condition progresses, the peripheral nerve myelin sheaths become inflamed and paralysis and sensory loss set in. When treated early in its course with immunosuppressant therapy such as plasmapheresis or intravenous gamma globulin, the damage can be minimized or even halted. If not treated early, the inflammation spreads beyond the nerve sheath and attacks the nerve axon itself. Once there is axonal damage, the prognosis worsens considerably. Peripheral nerve axons can regenerate one millimeter per day from the spinal cord down, taking up to a year to reinnervate the distal arms and legs. That is, if the nerves even grow back and manage to find their way. Sadly, it had been almost a year since Mia's symptoms began and there was no improvement.

I set Mia up right away for intravenous gamma globulin treatments, consisting of human antibodies from donated human plasma designed to block the inflammatory process that is destroying the myelin sheath. These treatments were scheduled for every three weeks, along with outpatient physical and occupational therapy, in addition to a nutrition consultation.

Now that Mia had a diagnosis, I was hopeful she would no longer be greeted with patronizing smiles and hollow reassurances. This sweet young woman in the prime of her life had not been taken seriously soon enough, and the cost to her quality of life was enormous. If her condition had been recognized and treated sooner, the destruction to her nerves may have been much less severe and possibly able to recover. Now that she had extensive axonal damage to her nerves, her prognosis for recovery was poor. The ongoing intravenous gamma globulin infusions would merely halt any further inflammation and damage since CIDP can be chronic.

Medical gaslighting is a very real phenomenon. And it is not restricted to male authority figures, as women physicians are guilty as well. Sadly, the patients most affected are those often characterized as complainers, such as women in general, the elderly, and the obese, as they are often discriminated against as someone who can't manage their own weight. It is well known that there can be emotional reasons for excessive bodily complaints, but everyone deserves the respect of being heard and having their complaints taken seriously, as there is often an underlying physical cause as well. Even when the cause is found to be emotional, isn't that something that should be addressed, as it is the cause of severe distress?

If you are unsure if you are being taken seriously, my advice is to trust your instincts.

The top signs of medical gaslighting include:

- They dismissed your symptoms.
- They invalidated your experience.
- They blamed or shamed you.
- They refused further testing or treatment.
- They ignored your medical history.
- They lacked empathy or compassion.
- They intimidated you.

CHAPTER 30

Thank You for Your Service

Today, as I celebrate the life of my beloved and loyal companion Prancer, a golden retriever therapy animal, I can't help but recall our life together. Well-loved and well-educated, she was so much more than a therapy animal. She was also my spirit animal, sent to me to provide guidance, protection, and strength. By my side for fourteen years, this gentle, devoted golden brought comfort, joy, and healing to countless people who needed it most. Her warm, loving spirit was an inspiration to all who were fortunate to encounter her. This gentle soul had an uncanny ability to sense when someone was in distress and offer a comforting presence. Always eager to please, she brought joy to many who were suffering from physical, emotional, or mental illness. Prancer touched more lives than we will ever know. Her impact on my life and the lives of my patients serves as a reminder that love is powerful, and that the healing power of compassion should never be underestimated.

Prancer came to me as an eight-week-old puppy Christmas present. I had lost another golden two years earlier and wasn't sure I was ready, but there she was. Arriving during a blizzard—hence the name Prancer—she captured my heart. I tried to leave her home with our older dog, but she wouldn't have it. She seemed to have a bad case of separation anxiety, and when she chewed up my wooden staircase, I became desperate and decided to take her to the office with me, at least temporarily. I foolishly thought I could keep her away from the patient area. After all, she had her very own love seat to lounge on with a view of the seagulls and a room full of toys. And I checked on her between patients and took her out to play every chance I got.

Prancer didn't take well to this idea either, howling for companionship when she was left for even ten minutes, alerting my patients that there was a dog on the premises. Not just any dog, but a golden retriever puppy! Naturally, they all wanted to meet her. Great, I thought. Just what I need. It was hard enough to try to stay on time with a demanding patient schedule. Now, playing with the puppy was going to use up even more time.

As if it was meant to be, Prancer immediately sensed that she and I both had an important job to do. Once I entered the examination room, she would gently scratch at the door to request entry. If this were granted, she would immediately plant herself next to the patient's chair, gratefully accepting pets when given but never disrupting the flow of the interview or examination. She was truly amazing! Before long, she was so requested by my patients that I didn't dare leave her at home.

As Prancer became an increasingly important part of my neurology practice, I knew it was just a matter of time before I would have to get certification for her if she was going to remain with me in any official capacity. I initiated the arduous process of getting us certified as a Pet Partner team, the only organization recognized by my hospital. I could check off the list of requirements easily: veterinary exam, updated vaccinations, obedience, and temperament. With all the paperwork complete, the two of us showed up for the official test before a series of judges. Prancer charmed them all, of course: she obeyed every voice and hand command without hesitation, sat patiently when she was rushed by groups of people, gently took a treat when offered, and didn't so much as flinch when a wheelchair bumped into her.

We had just one more task—the neutral dog test, where she would have to walk obediently by my side, paying no attention to a human with their therapy animal coming toward us. If we failed even one task, the outcome would be a failure, and there would be no do-overs. A therapy animal must be absolutely predictable to practice under the insurance umbrella of the Pet Partner program. I guess I neglected to explain that part to Prancer, for as friendly as she was, she couldn't resist happily greeting the other therapy animal as we approached. We had failed! Seeing my disappointment, Prancer knew right away she had done something terribly wrong.

The look in her eyes said it all. "What, Mom? You always praised me before when I greeted others enthusiastically. What did I do wrong?"

It took some doing, but I managed to convince the judges that I had not prepared Prancer adequately for that one station. They agreed to one do-over, but we would have to take the entire test again. So, off we went to practice walking around the lake daily, passing every type of dog and human, friendly and not. As hard as I tried, Prancer just couldn't be consistent with this, impulsively saying hello to anyone randomly.

The day of reckoning finally arrived, with the three judges peering at us over their glasses. "Are you two ready?"

"Please give us a moment," I pleaded. "I just need to have a talk with Prancer before we get started." I crouched down and, looking her straight in the eyes, said, "Please, Prancer, do this for me just this once and I will never ask you again. Walk past that dog without so much as a look her way. Our ability to continue to work together as a team depends on it."

Prancer performed perfectly for me that day, and we walked out with our joint Pet Partner certificate. We were now official and could visit patients together anywhere. Prancer never walked past another dog without saying hello to them again and, as promised, I never asked her to.

Patients were always given the choice as to whether they wanted Prancer to accompany us into the examination area, and almost all of them welcomed her. One day I was consulting with an elderly Vietnamese gentleman who initially said he was not sure. He granted her permission to enter but gestured for her to stay on the other side of the room and she dutifully obeyed. As the visit went on, however, I noticed that his eyes frequently strayed over to her. I shall never forget what he said to me when he left as he bent down to gently stroke her head.

"I noticed Prancer has a spotted tongue," he said. "In my culture that's a sign of magical healing powers in an animal. Take special care of her."

Then came Steven.

It was another routine day for Prancer and me at the office as we busily went from patient to patient. This morning, I was asked to consult on Steven, a young man paralyzed from the neck down because of a devastating spinal cord injury after a motorcycle crash. Fortunate to have survived, he was serving a life sentence in a wheelchair, requiring maximum assistance for all his needs. Accompanied by at least five family members, all of them spread out in the exam room on the stools, chairs, and exam table, Steven sat alone in his superpowered wheelchair in the center of the room, waiting to meet yet another doctor who would be unable to offer him any real assistance. He was a good-looking young man with blue eyes and chiseled features framing his sad countenance. His hands folded, lying motionless on his lap, with a green plaid throw over his shriveled legs. I imagined he was at least six feet tall, looking like he had once been athletic. He reminded me of my own son, Rory, who was just about his age, with a devilish personality, always pushing the limits with athletics and nature. Except for sheer luck, this could easily have been my son.

As I entered the room leaving the door open behind me, Steven politely acknowledged my presence, but he appeared despondent as he deferred the relaying of his history to his family in response to my queries. They explained that they had asked to see me in the hope that I could offer a comprehensive management plan to address Steven's complex needs, not the least of which was depression. As a mother myself, I could feel the pain of not only the patient but his family.

His injury was devastating and permanent. Knowing there was nothing to do but assist with rehabilitative needs such as spasticity, bowel, and bladder management, and make certain that they had all the social services they were entitled to, I struggled to find something more meaningful to offer them. In my desperation to offer anything of use for this young man, I had completely forgotten about Prancer, who was accustomed to being invited in after first scratching at the door. But today the door had been left open. As the visit was underway, Steven suddenly looked past me at the door. Flustered, I started to explain that I had forgotten to inform them there was a therapy dog on the premises, but he didn't seem to hear me. Prancer was just sitting in the doorway staring at Steven, not knowing what to do. It was Steven who finally broke the silence. "I love dogs so much," he said. "I would give anything . . . anything if I could only pet that dog."

At that moment, the room heavy with despair, Prancer, without a word from anyone or any signal from me, slowly rose, walked over to Steven, sat down next to his chair, and very gently placed her muzzle on the lifeless hands folded on his lap. There wasn't a dry eye in the room, including mine. Prancer remained there for the entire hour-long visit. She did more for that young man that day than I ever could.

There were days like this when spectacular things would happen, while many days were just routine. Despite this, you never wavered in your compassion, my dear Prancer. Only I knew what you were really thinking on those days when you would lie down with a heavy sigh, as if to say, "Oh no . . . not *another* headache!"

Day in and day out, year after year, you were my special companion, my teacher, and protector. You taught me the joy of unconditional love, the power of kindness and patience, and the importance of living in the present moment. You showed me the true meaning of loyalty and friendship, and that not all relationships need words.

Today, I remember and honor this amazing being and the incredible impact she had on so many lives. Goodbye, dear friend.

You are missed.

Conclusion

As I reflect on my journey through the world of neurology, I'm reminded that much of my career has been about listening to what cannot always be easily heard—the quiet signals, the subtle symptoms, the whispers of the mind. Neurology taught me that beneath the loud cries of disease, there are whispers that speak of resilience, hope, and the intricate connections between the body and the spirit.

Every patient I encountered offered me not just a clinical puzzle to solve but a chance to hear their whispered fears, hopes, and needs. In the rush of a busy practice, those whispers guided me, helping me see beyond the diagnoses to the human beings behind them. It is in those quiet moments, sitting with a patient and sharing a difficult truth, that I found the deepest meaning in my work.

Over the course of my career, I've witnessed minds falter, memories fade, and bodies weaken—but I have also witnessed incredible strength in the face of fragility. Through it all, I've learned that the mind, even in its silence, holds immeasurable power. It whispers stories of courage, endurance, and grace in the face of uncertainty.

Whispers of the Mind is not just the story of my patients; it is also the story of my own evolution—from a curious medical student to a neurologist who learned to listen deeply, not just to the brain, but to the soul. And in these whispers, I have found not just diagnoses but moments of profound humanity.

As I close this chapter of my career and my reflections, I realize that while I may have often been the one delivering difficult news, I was also receiving something in return—insight into the remarkable resilience of the human mind. I leave with gratitude for every patient who allowed me to walk with them through the mysteries and challenges of their own whispered journeys.

The mind, like life itself, is filled with whispers—soft, fleeting, sometimes unnoticed. My hope is that, through these essays, readers can hear those whispers, too, and in them, find stories not just of illness, but of strength, compassion, and connection.

Acknowledgments

First and foremost, my deepest gratitude to my family for their unwavering love, support, and patience. You have encouraged and believed in me throughout my journey in medicine and in writing this book.

I would also like to extend my heartfelt gratitude to my mentors and colleagues in neurology. Your guidance and insights have been invaluable, shaping not only my career but also the stories within these pages. I am privileged to have learned from you and to share in the incredible experiences that have enriched my professional and personal growth.

To my editor, Arnold Mann, your thoughtful expertise has been invaluable in helping to bring this book to life. I am deeply thankful for your efforts in refining every word, ensuring the essence of this work shines through.

Above all, I am profoundly grateful to my patients and their families. Your courage, resilience, and openness have been a wellspring of inspiration, teaching me invaluable lessons about life, strength, and the human spirit. It has been an honor to walk alongside you in your journeys, and I hope this book serves as a testament to your remarkable stories.

And, of course, to Prancer, my beloved therapy dog, who brought comfort and joy to my patients and inspired some of the stories in this book. Though she passed before I began writing, her gentle spirit and the solace she provided continue to echo through these pages. Prancer, with her head resting on patients' laps and her calm presence in the room, was a true heroine in our work together—a presence I carry with me and celebrate in every story she touched.

About the Author

Ken Taylor

Carolyn Larkin Taylor is a seasoned neurologist who has dedicated over three decades to the practice of medicine. A graduate of the University of Notre Dame and Hahnemann Medical College, she completed a neurology residency at the University of Pennsylvania, where she was awarded the Humaneness in Medicine Award and recognized as one of Philadelphia's Top Docs for Women. Her manuscript *Through a Mother's Eyes* was awarded second place for best unpublished memoir by the Pacific Northwest Writers Association. Dr. Taylor lives in Bellingham, Washington.

Looking for your next great read?

We can help!

Visit www.shewritespress.com/next-read
or scan the QR code below for a list
of our recommended titles.